NISTIR 7520

Diesel Adsorption to PVC and Iron During Contaminated Water Flow and Flushing Tests

Mark A. Kedzierski

U.S DEPARTMENT OF COMMERCE
National Institute of Standard and Technology
Building Environment Division
Building and Fire Research Laboratory
Gaithersburg, MD 20899-8631

June 2008

U.S. Department of Commerce
Carlos M. Gutierrez, Secretary

National Institute of Standards and Technology
James M. Turner, Acting Director

Diesel Adsorption to PVC and Iron During Contaminated Water Flow and Flushing Tests

M. A. Kedzierski
National Institute of Standards and Technology
Bldg. 226, Rm B114
Gaithersburg, MD 20899
Phone: (301) 975-5282
Fax: (301) 975-8973

ABSTRACT

This paper presents an experimental and theoretical study of aqueous diesel contamination and decontamination of a polyvinyl chloride (PVC) surface and an iron (Fe) surface. A test apparatus designed for the purpose of studying adsorption of diesel from a flowing dilute diesel/water mixture was used to measure the mass of diesel adsorbed per unit surface area (the excess surface density) and the bulk concentration of the diesel in the flow using a fluorescence based measurement technique. Both bulk composition and the excess surface density measurements were achieved via a traverse of the fluorescent measurement probe perpendicular to the test surface. The diesel adsorption to each test surface was examined for three different Reynolds numbers between zero and 7000. Measurements for a given condition were made over a period of approximately 200 h for a diesel mass fraction of approximately 0.15 % in tap water. For a Reynolds number of approximately 7000, the largest excess layer thickness was approximately 4.4 µm, which was measured on a PVC surface. Averaging over all contaminating flow rates and exposure times, the excess layer thickness on the PVC surface was approximately 2.0 µm. Reynolds number had little or no effect on the accumulation of diesel on an iron surface, which was approximately 0.71 µm. The adsorbed diesel on the PVC and iron surfaces was removed by flushing with tap water. Models to predict excess layer thickness during flushing and contamination were developed. The models predict flushing times to within 7 h and predict the influence of pipe surface on contamination level.

Keywords: adsorption, contaminant, diesel, excess layer, fluorescence, measurement technique, sorption, water

SECURITY NOTICE

THE MATERIAL IN THIS REPORT HAS NOT BEEN APPROVED FOR PUBLIC RELEASE, AND ITS USE IS RESTRICTED TO OFFICIAL DISTRIBUTION. WORK DESCRIBED IN THIS DOCUMENT MAY INVOLVE PROPRIETARY, NATIONAL SECURITY, OR OTHER SENSITIVE INFORMATION. INFORMATION OF THIS TYPE MAY ONLY BE RELEASED TO A PERSON WHO IS A UNITED STATES CITIZEN OR A PERMANENT RESIDENT ALIEN AND WHO IS AUTHORIZED TO RECEIVE THE INFORMATION. AUTHORIZATION MAY INCLUDE HAVING THE APPROPRIATE NATIONAL SECURITY CLEARANCE AND/OR SPECIFIC AUTHORITY TO RECEIVE SUCH INFORMATION. SPECIFIC APPROVAL BY DIVISION OR LABORATORY MANAGEMENT IS REQUIRED BEFORE ANY INFORMATION ARISING FROM THIS WORK IS DISCLOSED AS PER THE REQUIREMENTS OF THE NIST TECHNICAL COMMUNICATIONS PROGRAM (ADMIN MANUAL §4.09).

TABLE OF CONTENTS

ABSTRACT 1

LIST OF FIGURES 4

LIST OF TABLES 4

INTRODUCTION 5

EXPERIMENTAL APPARATUS AND UNCERTAINIES 5

TEST FLUIDS 7

MEASUREMENTS AND UNCERTAINTIES 7
    Fluorescence/Mass Calibration 7
    Flushing 10
    Air Gap Method 11

MEASUREMENT RESULTS 11
    Excess Layer Thickness 11
    Contamination Tests 12
    Flushing Tests 13
    Gap Test Check 14

MODEL DEVELOPMENT 14
    Flushing Model 14
    Contamination Model 16

FUTURE WORK 17

CONCLUSIONS 18

ACKNOWLEDGEMENTS 19

NOMENCLATURE 20
    English Symbols 20
    Greek symbols 20
    English Subscripts 20
    Superscripts 21

REFERENCES 22

APPENDIX A: UNCERTAINTIES 38

APPENDIX B: MEASUREMENT CORRECTION 42

APPENDIX C: MODEL DEVELOPMENT 44

    Flushing Model 44
    Contamination Model 46
APPENDIX D: FLUORESCENCE OF TOLUENE 49

LIST OF FIGURES

Fig. 1 Schematic of test loop 24
Fig. 2 Schematic of spectrofluorometer, test section, and linear positioning device 25
Fig. 3 Schematic of right angle spectrofluorometer 26
Fig. 4 Cross-sectional illustration of test section during contamination and flushing ... 27
Fig. 5 Schematic of fluorescence/composition calibration jar 28
Fig. 6 Overall calibration of Beer-Lambert Bougher law for diesel on copper disk 29
Fig. 7 Sample F vs. l fit for a PVC data set 30
Fig. 8 Effect of exposure time and flow rate on thickness of the diesel excess layer for a 0.15 % bulk freestream mass fraction on a PVC disk 31
Fig. 9 Effect of exposure time and flow rate on thickness of the diesel excess layer for a 0.15 % bulk freestream mass fraction on an iron disk 32
Fig. 10 Diesel excess layer thickness as a function of Re for PVC surface and water/diesel (99.85/0.15) 33
Fig. 11 Diesel excess layer thickness as a function of Re for iron surface and water/diesel (99.85/0.15) 34
Fig. 12 Flushing measurements used to fit coefficients of model 35
Fig. 13 Modified kinematic viscosity to account for adhesive forces between iron and diesel derived for two flow conditions (Re = 3200 and Re = 7000). 36
Fig. 14 Maximum contamination diesel layer on various pipe surfaces as predicted by eq. (22) 37
Fig. A.1 Relative uncertainty of l_e for 95 % confidence level for iron surface x_b = 0.15 % 38
Fig. A.2 Relative uncertainty of l_e for 95 % confidence level for PVC surface x_b = 0.15 % 39
Fig. A.3 Relative uncertainty of x_b for 95 % confidence level for iron surface 40
Fig. A.4 Relative uncertainty of x_b for 95 % confidence level for PVC surface 41
Fig. B.1 Basis for correction of iron surface measurements made after test apparatus repair 43
Fig. D.1 Schematic of experimental setup to measure fluorescence emission of toluene on three pipe surfaces: PVC, copper, and iron 49
Fig. D.2 Fluorescence emission of toluene on three pipe surfaces: PVC, copper, and iron 50

LIST OF TABLES

Table 1 Comparison of flushing model predictions to measurements 23

INTRODUCTION
Since the signing of the Executive Order establishing the Office of Homeland Security, Federal agencies have been working on ways to improve the security of the general public. In one example, the National Institute of Standards and Technology (NIST) is helping the U.S. Environmental Protection Agency (EPA) devise ways to safeguard the nation's drinking water supply. EPA is conducting potable water research with NIST on six different efforts. This report describes one of those efforts designed to fundamentally understand the attachment and detachment mechanisms of contaminants to solid plumbing materials under dynamic water flow conditions. The results of this work provide EPA with an investigative tool to support the development of a response to water contamination events and a potential detection technique for timely warning of such events.

The purpose of this study is to use a NIST fluorescence based measurement technique (Kedzierski, 2006) to add to the existing data on aqueous diesel adsorption to solid surfaces and to support the development of flushing and contamination models. In a previous study (Kedzierski, 2006), the diesel excess surface density was measured on an oxidized copper surface. The present study expands the database to diesel attachment to a polyvinyl chloride (PVC) surface and to an iron surface. In this way, we not only gain vital fundamental modeling information but we lay the groundwork for a possible early detection/monitoring system for sticky contaminants. These efforts have formed the foundation for future work that will focus on using the NIST water loop and the calibration technique to measure the accumulation and removal of diesel as a function of free-stream diesel concentration and contaminated water flow rate.

Commercial diesel was used rather than a chemically simpler surrogate in order to demonstrate the use of the technique with an actual potential contaminant. Diesel was also a desirable test contaminant because it has been found to exhibit a strong fluorescence. However, because of the complexity and the variability of diesel, the diesel for the project was restricted to a single batch. In this way, we can ensure the consistency of the properties of pure diesel[1] such as its liquid density and fluorescence characteristics.

EXPERIMENTAL APPARATUS AND UNCERTAINIES
The standard uncertainty (u_i) is the positive square root of the estimated variance u_i^2. The individual standard uncertainties are combined to obtain the expanded uncertainty (U), which is calculated from the law of propagation of uncertainty with a coverage factor. All measurement uncertainties are reported at the 95 % confidence level except where specified otherwise.

Figure 1 schematically shows the flow loop for measuring diesel on pipe substrates. The primary components of the loop are the pump, the reservoir, and the test chamber with the test section. The inside surfaces of the approximately 96 mm x 1.6 mm rectangular flow cross-section of the aluminum test chamber, shown in Fig. 2, were black anodized to

---

[1] "Pure diesel" is used here to denote that the particular batch of diesel, which will be consistently used throughout this project, is not mixed with water.

minimize stray light reflections. The channel was designed to have the same flow area as a 13 mm (nominally 0.5 inch) diameter copper pipe. The test chamber had a circular cavity to accept the solid pipe test surface. The height of the channel was 1.6 mm so that the probe could be flush to the top of the test section while maintaining proximity to the test surface for measurement purposes without being an obstruction to the flow. A centrifugal pump delivered the contaminated water to the entrance of the rectangular test chamber at room temperature. The pump head was removable so that it could be easily replaced in order to test a different contaminant. The flow rate was controlled and varied by varying the pump speed with a frequency inverter. A heat exchanger immersed in the reservoir was supplied with brine from a temperature-controlled bath to maintain the entrance temperature to the test chamber at ambient temperature (293.8 K). This was done to ensure that the diesel was at the same temperature as it was during the fluorescence calibration to avoid the temperature effect on fluorescence (Miller, 1981). An additional temperature-controlled bath was used to maintain the fluorescence standards at the same ambient temperature. As described in a succeeding section, the fluorescence standards were used to calibrate the range of the fluorescence measurements.

Residential copper pipe was used to plumb together the various components of the loop. Redundant volume flow rate measurements were made with an ultrasonic doppler and a turbine flowmeter with expanded uncertainties of $\pm$ 0.12 m^3/h and $\pm$ 0.03 m^3/h, respectively. As shown in Fig. 1, three water pressure taps before and after the test chamber permitted the measurement of the upstream absolute pressure and the pressure drops along the test section with expanded uncertainties of $\pm$ 0.24 kPa and $\pm$ 1.5 kPa, respectively. Also, a sheathed thermocouple measured the water temperature at each end of the test chamber to within an uncertainty of $\pm$ 0.25 K. The dissolved oxygen level, the conductivity, and the pH, were monitored at the water reservoir with associated B-type uncertainties of $\pm$ 0.5 %, $\pm$ 50 $\mu\Omega$/cm, and $\pm$ 0.3, respectively.

Figure 1 also shows the inlet and exit taps that were used to flush the test section with fresh tap water. In preparation for flushing, the test section was isolated from the rest of the test loop by closing valves. Then the fluid was drained from the test chamber and returned to the reservoir. Next, a tap water supply was connected to a test chamber port. The other test chamber port was connected to a filter to absorb any diesel before it was sent to a drain.

Figure 2 shows a view of the spectrofluorometer that was used to make the fluorescence measurements and the test chamber with the fluorescence probe perpendicular to the flattened pipe test surface. Figure 3 shows a simplified schematic of the right angle spectrofluorometer consisting of a xenon light source, an excitation and an emission monochromator, and an emission photomultiplier tube (detector). As sold off-the-shelf, the spectrofluorometer was designed to accept 45 mm × 10 mm × 10 mm fluorescent samples or cuvettes filled with fluorescent material. The spectrofluorometer was modified by replacing the cuvette holder with a special adapter with lenses and mirrors to remotely excite and measure fluorescence via a bifurcated optical bundle. Two optical bundles consisting of 84 fibers each originated from the spectrofluorometer. One of the bundles transmitted the excitation light, i.e., the incident intensity (I_o), to the test pipe surface. The other bundle carried the emission, i.e., the fluorescence intensity (F), from the test surface to the spectrofluorometer. The optical bundles originating from the spectrofluorometer merge transmitting and receiving fibers randomly into a single probe before entering the test section chamber. The sensor end of the fluorescence probe is sheathed with a quartz tube to protect it from reacting with the contaminant in the test fluid.

The excitation wavelength (λ_x) and the emission/detection wavelength (λ_m) were set to 434 nm and 485 nm, respectively, for all tests. Further details on the fluorescence measurement technique are given in Kedzierski (2006).

TEST FLUIDS
Number 2 diesel fuel was used from a single batch throughout the experiment to avoid property variations that might be caused by batch variations due to it being a complex mixture of hydrocarbons. Kedzierski (2006) provides the measured viscosity and density of the pure diesel liquid. A nominally 1.5 % by mass diesel mixture was prepared with local Gaithersburg, MD tap water for the exposure/flow rate tests. The measured dissolved oxygen level, the conductivity, and the pH, of the water at 24 °C before mixing with diesel were found to be, 86.4 %, 358 µΩ/cm, and 7.04, respectively.

Because diesel and water are not miscible, the suction inlet to the pump was designed to entrain diesel from the reservoir and into the water flow. As shown schematically in Fig.1, the opening of the pump suction line in the reservoir is situated approximately 10 mm below the liquid-air interface. The return flow entering the bottom of the reservoir ensured good flow mixing of any components of diesel that may have been hydrolyzed and settled to the reservoir bottom. Figure 4 depicts the colloidal flow within the test section and the fluorescent measurement probe above it for the contamination and decontamination test conditions. The size of the droplets in the dispersed flow is exaggerated for illustration purposes. Both test conditions are shown to have an excess layer thickness (l_e) of undiluted, potentially hydrolyzed diesel adsorbed to the test surface. Because the molar mass of the diesel is unknown, the surface excess density (Γ) is defined in this work on a mass basis as (Kedzierski, 2001):

$$\Gamma = l_e(\rho_d - \rho_b x_b) \tag{1}$$

The density of liquid diesel is ρ_d. The density of the flowing bulk mixture (ρ_b) is evaluated at the bulk mass fraction of the mixture (x_b). The surface excess density is roughly the mass of diesel attached per surface area. The Γ and l_e are the primary measurements of this study. The l_e is measured in the y-direction, as shown in Fig. 4, with the origin at the fluid-surface interface.

MEASUREMENTS AND UNCERTAINTIES
Fluorescence/Mass Calibration
Fluorescence as a means for detecting a contaminant has its advantages in that its absorption and fluorescence spectra are like a fingerprint that can be used in its

identification. Consequently, by isolating the wavelength of light that the contaminant fluoresces, its intensity can be used to identify its mass. Full detail of the fluorescence based measurement technique used in this study is given in Kedzierski (2006) and given in brief below.

Two different calibration methods had to be combined due to the additional complexity caused by immiscible liquids. Both calibration techniques were used to quantify different functional aspects of the Beer-Lambert-Bougher law (Amadeo et al., 1971), which forms the basis of the calibration equation. The first method was used to obtain the relationship between diesel composition and fluorescence intensity for a fixed light path length (fixed probe height above the test surface). The first method would have been sufficient had the bulk composition of the flow remained the same as it was charged in the reservoir. Due to the immiscibility of the two fluids, the bulk composition of the flow differs from that in the reservoir. As a result, a second method is necessary to determine both the contaminant mass fraction and the excess layer thickness. The second method that was developed in this study relies on a perpendicular traverse of the flow stream with the measurement probe. To achieve this, a linear positioning device with a graduated knob was adapted to the quartz tube as shown in Fig. 2. The second method (traverse method) is used to calibrate the effect of contaminant thickness (path length) and the proximity of incident intensity. The traverse method is essential for splitting the total measured fluorescent intensity into two components: that from the diesel on the test surface and that from the diesel in the bulk flow stream. In this way, the mass the diesel on the test surface and the composition of the fluid stream are determined.

Two standard jars (see Fig. 5) were used as references standards to set the lower (0) and upper (100) limits of the intensity signal on the spectrofluorometer for raw measurements made at the test section (F_r). A jar that contained only pure water was used to zero the intensity. A second jar that contained pure diesel was used to set the intensity on the spectrofluorometer to 100. All raw-measured intensities (F_r) were numerically normalized by the intensity from the zero-jar (F_0) and the maximum-jar (F_{100}) as:

$$F = \frac{F_r - F_0}{F_{100} - F_0} \qquad (2)$$

where the intensity of the contamination data was adjusted by no more than 0.3 % to account for the small (typically within ± 1 K) difference in temperature between the test section and the bath containing the maximum- and the zero-jars (Kedzierski, 2006). The maximum correction for the flushing data was approximately 1.5 %, which was larger than for the contamination measurements due to the colder temperature of the house tap water.

The linear form of the Beer-Lambert-Bougher law (Amadeo et al., 1971) shows that the measured fluorescence intensity is related to the incident light intensity (I_o), the extinction coefficient (ε), the concentration of the fluorescent diesel (c), the path length of light (l), and the quantum efficiency of the fluorescence (Φ) as:

$$F = 2.3 I_o \varepsilon c l \Phi \rightarrow [\varepsilon c l \leq 0.05] \tag{3}$$

The linear criteria for eq. (3) ($\varepsilon c l \leq 0.05$) is satisfied for 78 % of the calibration data, and the absorbance ($\varepsilon c l$) did not exceed 0.063 for all of the data. As a result, the calibration measurements shown in Fig. 6 gave: $2.3 I_o \Phi \varepsilon M_c^{-1} = 1.04735 \left[\text{m}^2 \text{kg}^{-1} \right] e^{-209.23 [\text{m}^{-1}] l}$ when fitted to eq. (3) (Kedzierski, 2006). Using the calibration and expressing the concentration in terms of the bulk mass fraction and the bulk liquid density gives the calibration of the fluorescence intensity in terms of the mass fraction and path length as:

$$F = \frac{2.3 I_o \Phi \varepsilon}{M_c} l x_b \rho_b = 1.04735 \left[\frac{\text{m}^2}{\text{kg}} \right] l x_b \rho_b e^{-209.23 [\text{m}^{-1}] l} \tag{4}$$

Note that the concentration of the fluorescent diesel has been replaced with the product of the bulk contaminant (diesel) mass fraction (x_b) and the density of the bulk mixture (ρ_b) divided by the molar mass of the contaminant (M_c). The mixture densities were calculated from a linear mass weighted basis of the pure fluid specific volumes.

Because the actual concentration of the diesel entrained in the water flow stream is unknown, eq. (4) cannot be directly used to obtain the excess layer of diesel on the pipe surface. The Γ and the l_e must be obtained from additional information that is obtained from a perpendicular traverse of the flow stream. As shown in Fig. 2, a linear positioning device with a graduated knob was used to traverse and locate the quartz tube relative to the test surface and thus measure the path length of the incident light through the fluid. Measurements of the fluorescence intensity (F) for various path lengths provided sufficient information for obtaining both the bulk mass fraction and the excess layer thickness. The methodology for this is explained in the following.

The total fluorescence signal (F) can be separated into three components along the path length while assuming a uniform bulk mass fraction. The total intensity is the sum of that contributed by the bulk concentration ($F_l(x_m = x_b)$) for the entire path length and that in the diesel excess layer ($F_{le}(x_m = 1)$) minus the intensity that would have been due to the bulk concentration but did not occur because it was displaced by the excess layer ($F_{le}(x_m = x_b)$)

$$F = F_l(x_m = x_b) - F_{l_e}(x_m = x_b) + F_{l_e}(x_m = 1) \tag{5}$$

Substitution of eq. (4) into the components of the above equation and grouping like terms gives:

$$F = 2.3 I_o \Phi \varepsilon M_c^{-1} \left[l x_b \rho_b - l_e x_b \rho_b + l_e \rho_d \right] \tag{6}$$

Here ρ_d is the density of liquid diesel.

For a given probe traverse, the only variable in eq. (6) is the path length. Consequently, eq. (6) can be arranged in terms of two regression constants for a single traverse:

$$F = (A_0 + A_1 l) e^{-209.23 [\text{m}^{-1}] l} \tag{7}$$

Figure 7 illustrates the fit of eq. (7) for a PVC data set at a Re of 7000.

Comparison of eqs. (6) and (7), yields the bulk mass fraction as:

$$x_b = \frac{A_1 e^{-209.23 \text{m}^{-1} l}}{\rho_d 2.3 I_o \Phi \varepsilon M_c^{-1}} = \frac{A_1}{1.04735 \left[\text{m}^2 \text{kg}^{-1} \right] \rho_d} \tag{8}$$

and the excess layer thickness as:

$$l_e = \frac{A_0}{\rho_d e^{209.23 \text{m}^{-1} l} 2.3 I_o \Phi \varepsilon M_c^{-1} - A_1} = \frac{A_0}{1.04735 \left[\text{m}^2 \text{kg}^{-1} \right] \rho_d - A_1} \tag{9}[2]$$

As shown in Appendix A, the average uncertainty of l_e for the measurements with the iron and the PVC surface for the contamination measurements was approximately ± 0.06 µm, and ± 0.1 µm, respectively. The average uncertainty of x_b was approximately ± 0.00008.

Flushing
The diesel bulk mass fraction of the tap water used during the flushing tests is zero. For flushing tests, eq. (8) produced a non-zero bulk mass fraction with a magnitude close to the uncertainty of the measurement, i.e., typically 0.008 % (80 ppm). An alternative approach for the flushing tests that forces the bulk mass fraction to be zero is to start by setting $x_b = 0$ in eq. (6) and taking its derivative with respect to the path length.

---

[2] The methodology presented here is a refinement of that given in Kedzierski (2006). Equations (8) and (9) are more explicit than those presented in Kedzierski (2006), but give the same results.

Rearranging the resulting differentiated equation and solving for the excess layer thickness yields:

$$l_e = \frac{\dfrac{dF}{dl}}{209.23[\text{m}^{-1}]\rho_d 2.3 I_o \Phi \varepsilon M_c^{-1}} = \frac{A_0 - A_1\left(\dfrac{1}{209.23[\text{m}^{-1}]} - l\right)}{1.04735\left[\text{m}^2\text{kg}^{-1}\right]\rho_d} \tag{10}$$

Equations (9) and (10) are equivalent for negligible A_1, which is the case for flushing. However, eq. (10) was used to obtain the l_e for all of the flushing measurements because of its more explicit derivation. The average value of l_e was used for a given measurement probe traverse. As shown in Appendix A, the average uncertainty in l_e for the iron and the PVC flushing tests was approximately ± 0.02 μm, and ± 0.05 μm, respectively.

Air Gap Method
A secondary methodology was developed that relies on the gradient of F rather than its absolute value in order to confirm the measurement of l_e as obtained from eq. (9). The advantage of a gradient approach would be the elimination of a bias error on the measurement of F if it existed. As shown in Fig. 4, part of the excitation is reflected from the diesel-air interface and is not available to induce fluorescence. Consequently, the calibration must be adjusted to account for the air gap during the drained test chamber measurements. Kedzierski (2006) provides the derivation of the air-gap l_e and the result is given here as:

$$l_e = \frac{-0.0121[\text{m}]\dfrac{dF_{ag}}{dl}M_c}{2.3 I_o \Phi \varepsilon x_m \rho_m} = -0.01156\left[\frac{\text{kg}}{\text{m}}\right]\frac{\dfrac{dF_{ag}}{dl}e^{209.23[\text{m}^{-1}]l}}{x_m \rho_m} \tag{11}$$

MEASUREMENT RESULTS
Excess Layer Thickness
The test apparatus shown in Fig. 1 was used to submit either a PVC disk or an iron disk to exposure tests with fixed bulk concentration of diesel in tap water under varying flow conditions. More specifically, contamination measurements over an approximate 200 h time period were made for three different Reynolds numbers varying from 0 to 7000:

$$\text{Re} = \frac{4\dot{m}}{\mu_b p_w} \tag{12}$$

where the wetted perimeter of the channel was 195 mm, the viscosity of the mixed bulk flow (μ_b) was calculated using a nonlinear mixture equation, and the mass flow rate ($\dot{m}$) was obtained from the turbine meter. Flushing measurements were done for a fixed Re of approximately 5000. The range of Reynolds numbers result from using a range of volume flow rates that a half-inch diameter tube would experience in typical buildings.

After each contamination tests, the test surface was cleaned with acetone and clean tap water.

Contamination Tests

Figure 8 provides the measured diesel layer thickness on a PVC surface as caused by an exposure to a flowing water/diesel (99.85/0.15) mixture, i.e., diesel at approximately 0.15 % bulk mass fraction (1500 ppm). The exposure time is the duration of the exposure test: it is the time that the test surface is exposed to the contaminated flow starting with a clean surface. The open circle, square, and triangle symbols represent contamination measurements obtained from eq. (9) for the Re = 0, 3200, and 7000 conditions, respectively.

Figure 8 shows that the Re = 3200 and the soak (Re = 0) contamination tests gave similar results. More specifically, the excess layer thickness was established immediately upon exposure of the PVC surface to the water/diesel mixture and remained nearly constant for the 200 h test duration. Only a marginal increase in the time-averaged l_e was observed from approximately 1.32 µm to 1.48 µm when the Re was increased from 0 to 3200, respectively. However, an increase in the Re to 7000 resulted in roughly a 142 % increase in the time-averaged l_e over the soak condition to an average value of approximately 3.2 µm. In addition, the Re = 7000 condition did not produce a nearly constant l_e with respect to exposure time as did the Re = 0 and Re = 3200 conditions. Rather, the Re = 7000 condition gave a maximum diesel thickness of approximately 4.4 µm at an exposure time of 20 h. With further exposure to 140 h, the diesel thickness decreased from this maximum to nearly the thickness at the initial contamination, which was approximately 3 µm. For the PVC surface, the approximate average l_e for the Re = 0, 3200, and 7000 conditions was 1.32 µm, 1.48 µm, and 3.16 µm, respectively. Averaging over all contaminating flow rates and exposure times, the average l_e for x_b = 0.15 % on the PVC surface was approximately 2.0 µm.

Figures 9 gives the measured diesel layer thickness[3] on an iron surface due to exposure to a flowing water/diesel (99.85/0.15) mixture, i.e., the same composition as for the PVC surface. In general, the flow rate had little effect on the diesel excess layer thickness. For the iron surface, the approximate average l_e for the Re = 0, 3200, and 7000 conditions was 0.87 µm, 0.66 µm, and 0.60 µm, respectively. Averaging over all contaminating flow rates and exposure times, the average l_e for x_b = 0.15 % on the iron surface was approximately 0.71 µm. Consequently, the PVC surface adsorbs approximately 180 % more diesel than the iron surface. The average accumulation of diesel on a copper surface for x_b = 0.2 % was comparable to the PVC surface being approximately 2.3 µm (Kedzierski, 2006).

Figure 10 crossplots all of the contamination excess layer measurements of Fig. 8 as a function of Re. Figure 10 shows that the maximum diesel excess layer thickness on the

---

[3] The soak measurements on the iron surface were corrected as outline in Appendix B To account for additional rust resulting when the surface was exposed to air during repair of the apparatus.

PVC surface of approximately 4.4 μm occurred between Re of 5700 and 6300. The peak l_e for Re near 3200 was approximately 2 μm, which is approximately 55 % less than the maximum l_e for the Re = 6000 tests. Another 20 % reduction in the peak l_e on the PVC surface was observed when the nominal Re was reduced from 3200 to 0. The peak l_e for the soak tests (Re = 0) was approximately 1.6 μm for the PVC surface.

Figure 11 crossplots all of the excess layer measurements on the iron surface (Fig. 9) as a function of Re. Figure 11 shows that the maximum film thickness of approximately 1.3 μm occurred for the soak tests on the iron surface. The peak l_e on the iron surface decreases slightly from the soak condition for increasing Re. The peak l_e on the iron surface for Re near 3200 and 6000 was approximately 1.1 μm and 1.0 μm, respectively. The dashed lines given in Figs. 10 and 11 indicate the maximum measured excess layer for tests on the PVC and the iron surface. The variation in Re for a given set of tests for "fixed" Re was caused by an approximate 1 % variation in the water temperature during startup and the an approximate 15 % variation in the water flow during the nearly 200 h test duration.

Flushing Tests
The flushing tests that were done after each contamination test are shown in Figs. 8 and 9. Measurements of l_e during flushing of the surface after the Re = 0, 3200, and 7000 contamination tests are represented by the filled circle, square, and triangle symbols, respectively. For the PVC surface, most of the flushing measurements are close to but less than zero. The average of all the flushing measurements on the PVC surface is approximately –0.2 μm. It is likely that an unknown bias error has caused the measurement to be less than zero because 0.2 μm is larger than the uncertainty of the l_e measurement. The negative thicknesses are interpreted as a clean surface. Consequently, the surface is clean nearly immediately after the inception of flushing. Negative thickness were observed with the exception of the flushing tests after the Re = 7000 contamination (of the PVC surface) where the initial l_e was about 0.6 μm. The l_e decreased from approximately 0.6 μm to approximately 0.13 μm after flushing for approximately 3.6 h. This corresponds roughly to a 0.13 μm/h removal rate, which is similar in magnitude to the flushing diesel removal rate, 0.10 μm/h, found for a copper surface (Kedzierski, 2006).

The l_e averaged over all exposure times and Re for the flushing test on the iron surface was roughly 0.44 mm. The flushing tests after the Re = 7000 contamination were the most successful in getting the surface clean ($\overline{l_e} = -0.2 \cdot \mu m$) assuming that negative thickness beyond the uncertainty implies a clean surface. On average, the flushing after the Re = 0 and the Re = 3200 tests left similar quantities of diesel on the surface, 0.6 μm and 0.8 μm, respectively. These numbers indicated that flushing after the Re = 3200 contamination tests did not show any diesel removal, while the flushing after the soak tests gave only roughly a 0.3 μm diesel removal. However, both flushing tests after Re = 0 and the Re = 3200 contamination tests were corrected as outlined in Appendix B. The corrected measurements have an estimated uncertainty of ± 0.8 mm. Consequently, it is likely that the iron surfaces were flushed clean after Re = 0 and the Re = 3200 contamination tests on the iron surface given that the measurements are within the measurement uncertainty and that the non-corrected flushing measurements indicated that the iron surface was clean after flushing.

Gap Test Check
Filled symbols in Figs. 8 and 9, shown between 150 h and 200 h, represent l_e measurements that were made at the end of the exposure tests after the test section was drained using the air-gap technique as a secondary measurement technique. The average uncertainty of the l_e obtained from the air-gap measurement was approximately ± 400 % of the measurement. As a result, it is not surprising that the l_e obtained from the air-gap measurement and that from eq. (9) or eq. (10) ranged between being 300 % smaller and 200 % larger. It is not clear why the air-gap check gave good agreement with the measurements presented in Kedzierski (2006) but failed to provide confirmation of the present measurements.

MODEL DEVELOPMENT
The derivations of the models to predict the thickness of the contaminant excess layer on plumbing surfaces for the contamination and the flushing conditions are presented in Appendix C. Each model assumes that transfer of mass to and from the surface occurs solely in a direction that is perpendicular to the surface, i.e., in the y-direction shown in Fig. 4. The flushing model predicts the thickness of the excess layer as a function of time, contaminant transport properties, and flushing Reynolds number. The contamination model gives the maximum contaminant excess layer thickness that can occur for a given contaminant bulk mass fraction and surface affinity.

Flushing Model
The model for flushing with contaminant-free water is based on the conservation of contaminant mass within the excess layer (l_e). Because the excess layer is thin, it is approximated as stagnate in the axial direction such that the net motion of diesel is solely perpendicular from the surface in the y-direction. The model is governed by turbulent convection and diffusion of contaminant from the surface. The effect of diffusion is modeled by the ratio of the diffusion coefficient (D_{wd}) to the transition depth (B_T) over which the concentration difference occurs. The turbulent convection is modeled by the contaminant viscosity (ν_d), the friction velocity (u_*), and an entrainment constant (K_J) that relates the average entrainment velocity ($v'_{max}/2$) to the local axial-velocity (u) in the viscous sublayer as:

$$K_J \equiv v'_{max}/u \tag{13}$$

The resulting equation to predict the contaminant excess layer (l_e) as a function of flushing time (t) and initial excess layer thickness (l_{e0}) is:

$$\frac{l_e}{l_{e0}} = \left(1 + \frac{1}{K_D}\right) e^{\frac{-K_J u_*^2}{2\nu_d}t} - \frac{1}{K_D} \tag{14}$$

where the dimensionless constant K_D is a ratio of the convective to the diffusive influences:

$$K_D = \frac{K_J u_*^2 B_T l_{e0}}{2 v_d D_{dw}} \tag{15}$$

For $K_D = 1$, convection and diffusion fluxes of contaminant from the surface are equal at the beginning of the flushing. Values of K_D larger than 1 indicate that convection is more important than diffusion of contaminant from the surface.

The friction velocity is calculated from an equation given by Kays and Crawford (1980) with the average (bulk) axial velocity (and fluid properties) of the flushing water (V) and its Reynolds number as:

$$u_* = V\sqrt{0.039\,\mathrm{Re}^{-0.25}} \tag{16}$$

Flushing measurements for contamination levels larger than those of the scope of the present project were taken in order to establish changes in the measured excess layers that were large enough to fit to eq. (14). These experiments flushed diesel from a PVC surface with tap water flowing at an Re of approximately 5000. The initial diesel excess layer was approximately 31.4 μm. Figure 12 shows the flushing measurements that were used to obtain the K_J and D_{dw}/B_T constants from a least squares regression of the Fig. 12 data to eq. (14). The regression constants are:

$$K_J = 0.66 \times 10^{-8} \pm 0.05 \times 10^{-8} \tag{17}$$

$$\frac{D_{dw}}{B_T} = 0.6 \times 10^{-10} \left[\frac{m}{s}\right] \pm 0.1 \times 10^{-10} \left[\frac{m}{s}\right] \tag{18}$$

Fig. 12 plots eq. (14) using the regressed eq. (17) and (18) constants. The model predictions are within ± 4 μm of all the measured l_e.

The entrainment constant, K_J, is expected to be less of a function of the properties of the contaminant/flushing pair than is the diffusion constant. The ratio of the axial bulk velocity to its peak fluctuating component may be nearly constant because the bulk velocity is the potential for the fluctuating velocity. Considering this and that Kays and Crawford (1980) show that the ratio of the peak fluctuation turbulent velocity components is constant, the K_J may be relatively constant for a particular flushing fluid and for various Re. It is expected that the K_J presented here would be valid for water and liquids with similar kinematic viscosity.[4] Conversely, the D_{dw}/B_T depends on the properties of both the contaminant and the flushing fluid.

---

[4] However, K_J may be altered by the adhesion forces that must be overcome to remove the contaminant from the wall.

The required time (t) to flush the surface clean is derived in Appendix C and is presented here:

$$t = \frac{-2v_d}{K_J u_*^2} \ln\left[\frac{1}{K_D + 1}\right] \qquad (19)$$

Table 1 compares predictions using eq. (19) with observed values. All of the flushing times are predicted to within 7 h and all but one prediction is conservatively overestimated.

Contamination Model
For the contamination case, where a balance between deposition of contaminant on the surface and removal of the contaminant from the surface must be achieved, surface adhesion requires that a distinction be made between the velocity of the contaminant toward the surface (v_i) and that away from it (v_o). Very near the wall there is an additional resistance to flow away from the surface as caused by the attraction of diesel molecules to the molecules of the pipe surface. Likewise, for the same region near the pipe, the attractive forces induce a reduction in the resistance of flow toward the surface. Considering that it is flow that is being modeled, a simply way to approximate this behavior is via a modified viscosity. For example, the entrainment velocity (evaluated using the properties of the contaminated flow) given in Appendix C (eq. C8) for flow approaching the pipe surface becomes:

$$[v_i]_{y=l_e} = \frac{K_J u_*^2 l_e}{2v_d \tanh\left(\dfrac{l_e}{\delta}\right)} \qquad (20)$$

Here, the kinematic viscosity is less than the constant property viscosity (v_d) for distances from the surface less than the penetration depth, δ, because adhesive forces assist the flow of contaminant to the surface within this region. The magnitude of the penetration depth is determined by the affinity of the contaminant for the pipe surface. Consequently, each contaminant\pipe combination will have its own value of δ. The hyperbolic tangent was chosen for its simplicity and because it closely matched what was believed to be the required relationship with respect to l_e.

Likewise, surface adhesion acts to deter the flow of contaminant from the wall according to an assumed hyperbolic cotangent relationship with respect to l_e. Here the leaving velocity is approximated as a local increase in viscosity above the constant property value as the pipe surface is approached for distances less than δ:

$$[v_o]_{y=l_e} = \frac{K_J u_*^2 l_e}{2v_d \coth\left(\dfrac{l_e}{\delta}\right)} \qquad (21)$$

Figure 13 provides an example of the ratio of the adhesive influenced viscosity to the constant property viscosity of diesel (ϑ) as a function of the distance from the iron surface. For this example, the excess layer is 0.5 μm and the penetration depth is 14 μm. The viscosity ratio is plotted as a function of y to illustrate the waning and waxing influences of the adhesive forces between iron and diesel as a model concept. However, as far as the contamination model is concerned, ϑ is evaluated only at $l_e = 0.5$ μm.

Considering that the exposure time of a pipe surface to a contaminant may not always be known, a conservative decontamination response would be to flush for the maximum contaminant film thickness (excess layer) for the given concentration of contaminant in the flow. The maximum contaminant excess layer, which occurs at steady state where the contaminant deposition balances the removal at the wall, can be determined by setting the partial derivative of the excess layer with respect to time to zero (see Appendix C) and solving the resulting equation for l_e, yielding:

$$[l_e]_{max} = \frac{\delta}{2} \ln\left(\frac{1+\sqrt{x_b}}{1-\sqrt{x_b}}\right) \qquad (22)$$

Equation (22) can be used to determine the maximum possible contamination level and used as input to eq. (19) in order to calculate the required flushing time to obtain a clean surface. Mathematically, eq. (22) is valid for values of x_b between zero and 1; however, it has been validated for only dilute solutions.

Average values of the penetration depth were found by back-substituting the measured diesel bulk mass fraction and the measured $[l_e]_{max}$ into eq. (22) and solving for δ. The δ was found to be surface dependent: 150 μm, 42 μm, and 16 μm for copper, PVC, and iron, respectively. Figure 14 shows the maximum contamination layer for the three pipe surfaces as a function of x_b as predicted by eq. (22). Because of the presumed stronger affinity between copper and diesel, the copper surface has greater contamination levels for a given x_b as compared to the PVC and iron surfaces. The greater affinity is modeled via the larger penetration depth.

FUTURE WORK
The overall goal of this project is to develop the tools that are necessary for adequately responding to a contamination event. These tools have been envisioned to consist of measured data, predictive models, and a computer program that embodies the data and models. The software would allow the user to input information that is specific to a contamination event and to receive estimates of contamination levels and required flushing times. Such an all-encompassing goal requires a significantly large effort in order to ensure that the software makes reliable predictions for all possible contaminants. In order to achieve this task in a reasonably short time, the model must be physically based, but yet simple, so that it may be easily adjusted and extrapolated for different contaminants.

In this light, the models developed in this project require additional enhancement to ensure that the best possible predictions can be provided. Possibly, the major effort toward this end will be using measured data to calculate the constants K_J, D_{dw}/B_T, and δ for potential contaminants (like toluene, see Appendix D) and pipe surfaces. In the course of this effort, it will be determined whether or not K_J is merely a function of the properties of the flushing fluid. In addition, it can be determined if δ is constant for a particular contaminant/surface pair. In the short term, more flushing measurements at different initial l_e are required to validate the flushing constants considering that only one data set was used to obtain them. It may be appropriate to lump the diffusion that occurs during contamination into the adhesion effect. On the other hand, inclusion of a diffusion component in the contamination model may be necessary. These questions can be answered with further experimentation and model enhancement.

CONCLUSIONS

A detailed account of the continued development of a fluorescence based measurement technique for measuring the mass of contaminant on solid surfaces in the presence of water flow has been provided. A test apparatus was designed and developed to use the fluorescent properties of diesel for the purpose of studying its adsorption and desorption to and from plumbing pipe materials. A calibration technique was developed to measure both the mass of diesel adsorbed per unit surface area (the excess surface density) and the bulk concentration of the diesel in the flow.

Measurements for a given condition were made over a period of approximately 200 h for a diesel mass fraction of approximately 0.15 %. The largest excess layer thickness was approximately 4.4 μm, which was measured on a PVC surface for a Reynolds number of approximately 7000. The peak contaminant thickness on the PVC surface was shown to increase for increasing Re. Averaging over all contaminating flow rates and exposure times, the excess layer thickness on the PVC surface was approximately 2.0 μm. Reynolds number had little or no effect on the accumulation of diesel on an iron surface, which was approximately 0.71 μm. On average, the PVC surface adsorbed approximately 180 % more diesel than the iron surface. Both the PVC and iron surfaces were readily flushed with tap water.

Models to predict excess layer thickness on plumbing surfaces during flushing and contamination were developed. The flushing model predicts the thickness of the excess layer as a function of time, contaminant transport properties, and flushing Reynolds number. The required time to flush clean given an initial contaminant thickness can also be calculated. All of the flushing times were predicted to within 7 h. The contamination model gives the maximum contaminant excess layer thickness that can occur for a given contaminant bulk mass fraction and surface affinity. The contamination model accounts for the relative affinity between pipe surface and contaminant by predicting greater contamination levels on the PVC than on the iron surfaces. Together, the two models can be used as tools to develop a response to a contamination event.

ACKNOWLEDGEMENTS

This work was funded by the U.S. Environmental Protection Agency (EPA) under contract #DW-13-92167701-0, with the guidance of project manager Mr. V. Gallardo. Thanks go to the following NIST personnel for their constructive criticism of the first draft of the manuscript: Dr. S. Treado, and Dr. P. Domanski. Thanks goes to Dr. K. Cole of NIST's Biochemical Science Division for his constructive criticism of the second draft of the manuscript. Furthermore, the author extends appreciation to Mr. D. Wilmering and Mr. J. Wamsley for taking the measurements and conquering the difficult machining and design problems encountered in the project.

NOMENCLATURE

English Symbols

A	regression constants in eq. (7)
B_T	diffusion length, m
c	concentration, mol m^{-3}
$\bar{c}$	mass concentration, kg m^{-3}
D_{wd}	diffusion coefficient, m^2 s^{-1}
F	fluorescence intensity
F_c	corrected fluorescence intensity measurement
F_r	raw fluorescence intensity measurement
I_o	incident intensity, V
K_D	dimensionless number convection to diffusion eq. (15)
K_J	entrainment velocity ratio eq. (13)
l	path length, m
l_e	thickness of excess layer, m
M	mass of excess layer, kg
M_c	molar mass of contaminant, kg mol^{-1}
$\dot{m}$	mass flow rate, kg s^{-1}
Re	Reynolds number
p_w	wetted perimeter of channel, m
t	time, s
T	temperature, K
u	local axial velocity, m s^{-1}
u_*	friction velocity, m s^{-1}
U	expanded uncertainty
u_i	standard uncertainty
v	local velocity in y-direction, m s^{-1}
v'	turbulent fluctuating velocity in y-direction, m s^{-1}
V	bulk axial velocity, m s^{-1}
x	mass fraction of diesel
y	coordinate perpendicular to pipe surface, m

Greek symbols

δ	penetration depth, m
Γ	surface excess density, kg m^{-2}
ε	extinction coefficient
ϑ	ratio of modified to constant viscosity
λ	wavelength, m
μ	dynamic viscosity, kg m^{-1} s^{-1}
ν	kinematic viscosity, m^2 s^{-1}
ρ	mass density of liquid, kg m^{-3}
Φ	quantum efficiency of fluorescence

English Subscripts

0	zero reference jar
100	maximum reference jar

a	ambient
ag	air gap
b	bulk
d	pure diesel
e	excess layer
i	inlet
l	liquid
l_e	excess layer
m	emission, mixture
ng	no air gap
o	outlet or exit
s	solid surface
T_b	reference bath temperature
T_T	test section temperature
x	excitation
w	tap water

Superscripts

-	average

REFERENCES

Amadeo, J. P., Rosén C., and Pasby, T. L., 1971, Fluorescence Spectroscopy An Introduction for Biology and Medicine, Marcel Dekker, Inc., New York, p. 153.

Guilbault, G. G., 1967, Fluorescence: Theory, Instrumentation, and Practice, Edward Arnold LTD., London, pp. 91-95.

Herman, B., 1998, Fluorescence Microscopy, 2nd ed., Springer-Verlag New York, Inc., pp. 69 –71.

Kays, W. M., and Crawford, M. E., 1980, Convective Heat and Mass Transfer, McGraw-Hill, New York, NY.

Kedzierski, M. A, 2006, "Development of a Fluorescence Based Measurement Technique to Quantify Water Contaminants at Pipe Surfaces During Flow," NISTIR 7355, U.S. Department of Commerce, Washington, D.C.

Kedzierski, M. A., 2003, "Effect of Bulk Lubricant Concentration on the Excess Surface Density During R134a Pool Boiling with Extensive Measurement and Analysis Details," NISTIR 7051, U.S. Department of Commerce, Washington, D.C.

Kedzierski, M. A., 2002, "Use of Fluorescence to Measure the Lubricant Excess Surface Density During Pool Boiling," Int. J. Refrigeration, Vol. 25, pp.1110-1122.

Kedzierski, M. A., 2001, "Use of Fluorescence to Measure the Lubricant Excess Surface Density During Pool Boiling," NISTIR 6727, U.S. Department of Commerce, Washington, D.C.

Miller, J. N., 1981, Volume Two Standards in Fluorescence Spectrometry, Chapman and Hall, London, pp. 44-67.

Reader, J, Corliss, C. H., Wiese, W. L., and Martin, G. A., 1980, "Wavelengths and Transition Probabilities of Atoms and Atomic Ions", NSRDS-National Bureau of Standards #68, U.S. Department of Commerce, Washington.

Schlichting, H., 1979, Boundary-Layer Theory, McGraw-Hill, New York, pg. 591.

Schwarzenbach, R. P., Gschwend, P. M., and Imboden, D., M., 2003, Environmental Organic Chemistry, 2nd ed., Wiley, NJ, pp 281-283.

Simplex, 2006, http://www.simplexdirect.com/FuelSupply/mainten04.html

Tennekes, H., and Lumley, J. L., 1982, A First Course in Turbulence, MIT Press, Cambridge, MA, p. 160.

Table 1 Comparison of flushing model predictions to measurements

Surface	Re	l_{e0} (μm)	K_D	Flushing time for clean eq. (19) (h)	Observed flushing time for no change (h)
PVC	0	1.5	0.29	6.4	≈ 0
PVC	3200	1.5	0.29	6.4	≈ 0
PVC	7000	2.5	0.49	9.7	≈ 8
Iron	0	1.0	0.20	4.3	≈ 0
Iron	3200	1.0	0.20	4.3	≈ 0
Iron	7000	0.5	0.10	2.3	≈ 5

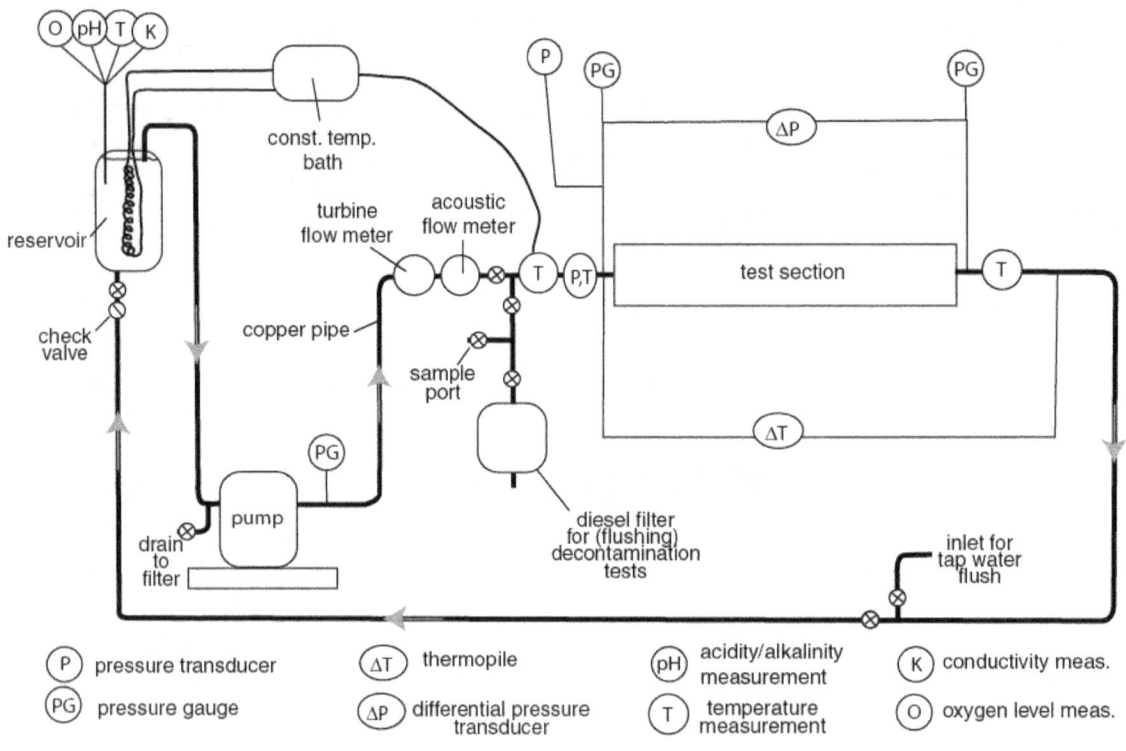

Fig. 1 Schematic of test loop

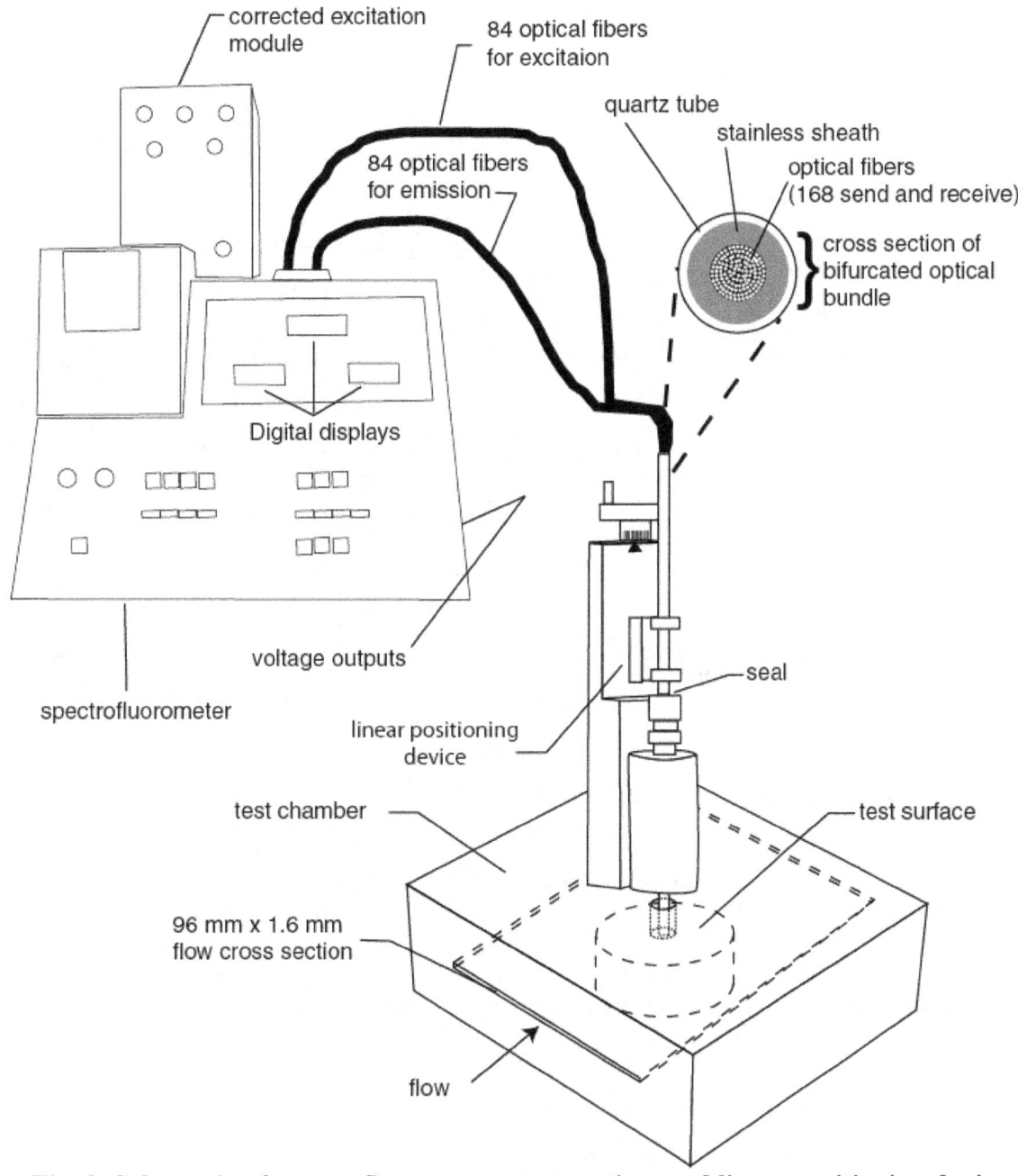

Fig. 2 Schematic of spectrofluorometer, test section, and linear positioning device

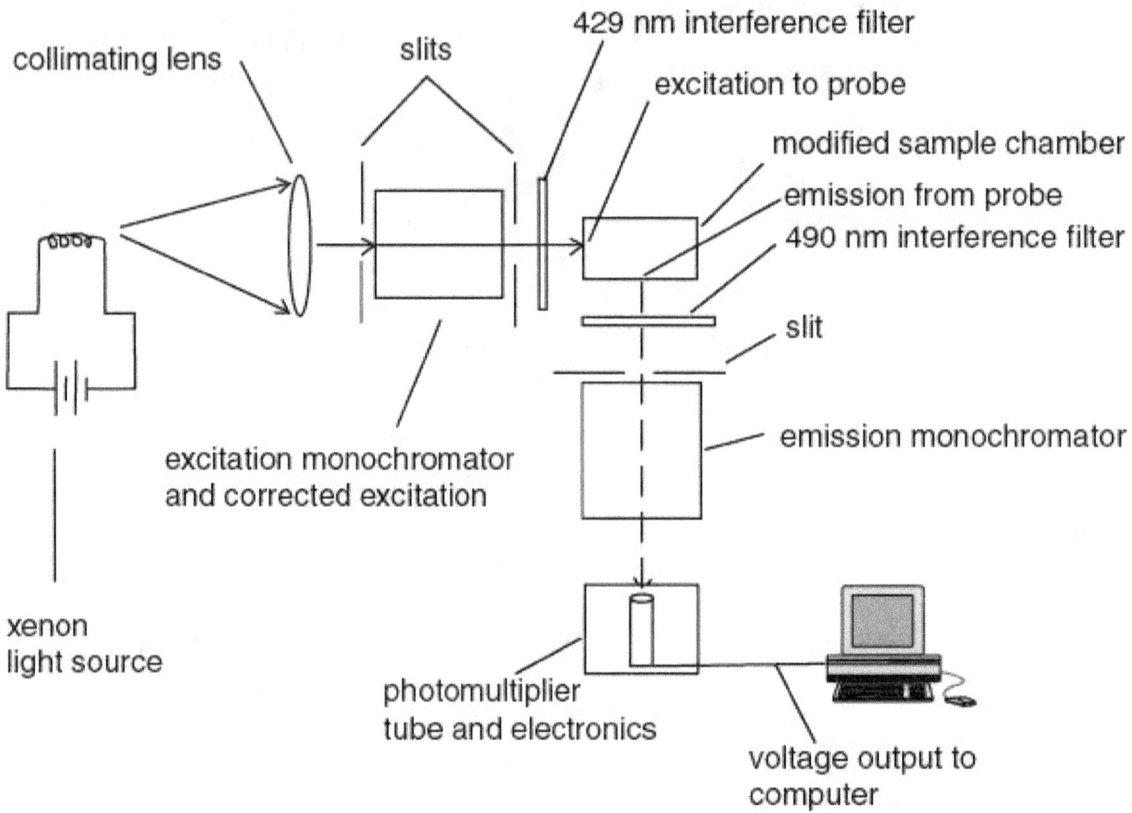

Fig. 3 Schematic of right angle spectrofluorometer (Kedzierski, 2006)

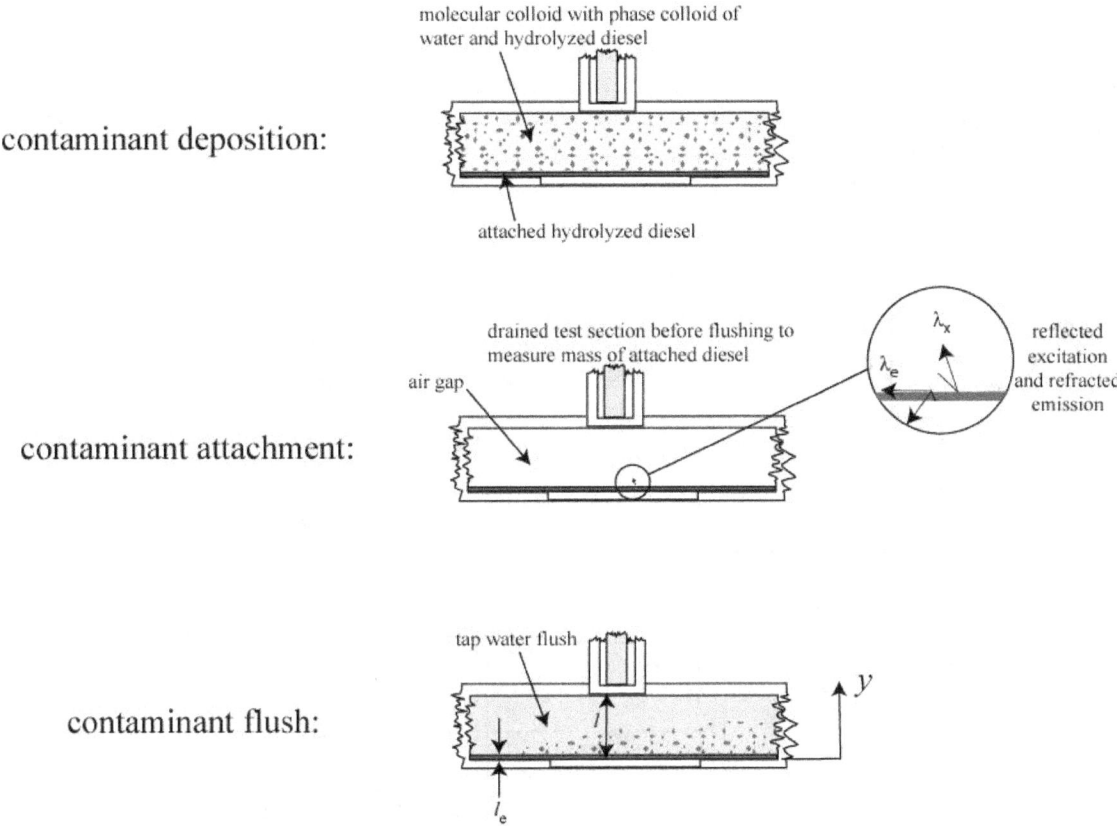

Fig. 4 Cross-sectional illustration of test section during contamination and flushing

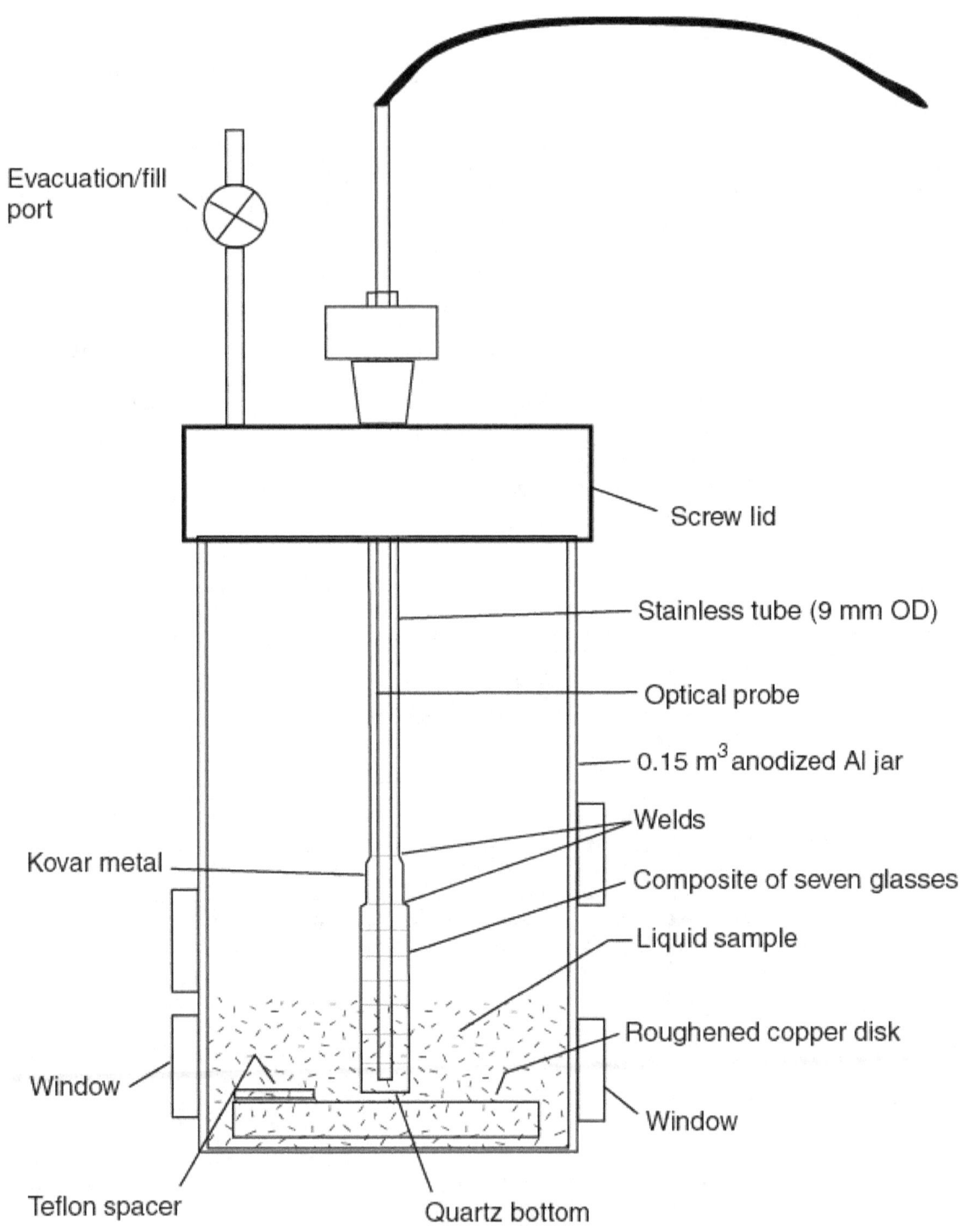

Fig. 5 Schematic of fluorescence/composition calibration jar (Kedzierski, 2006)

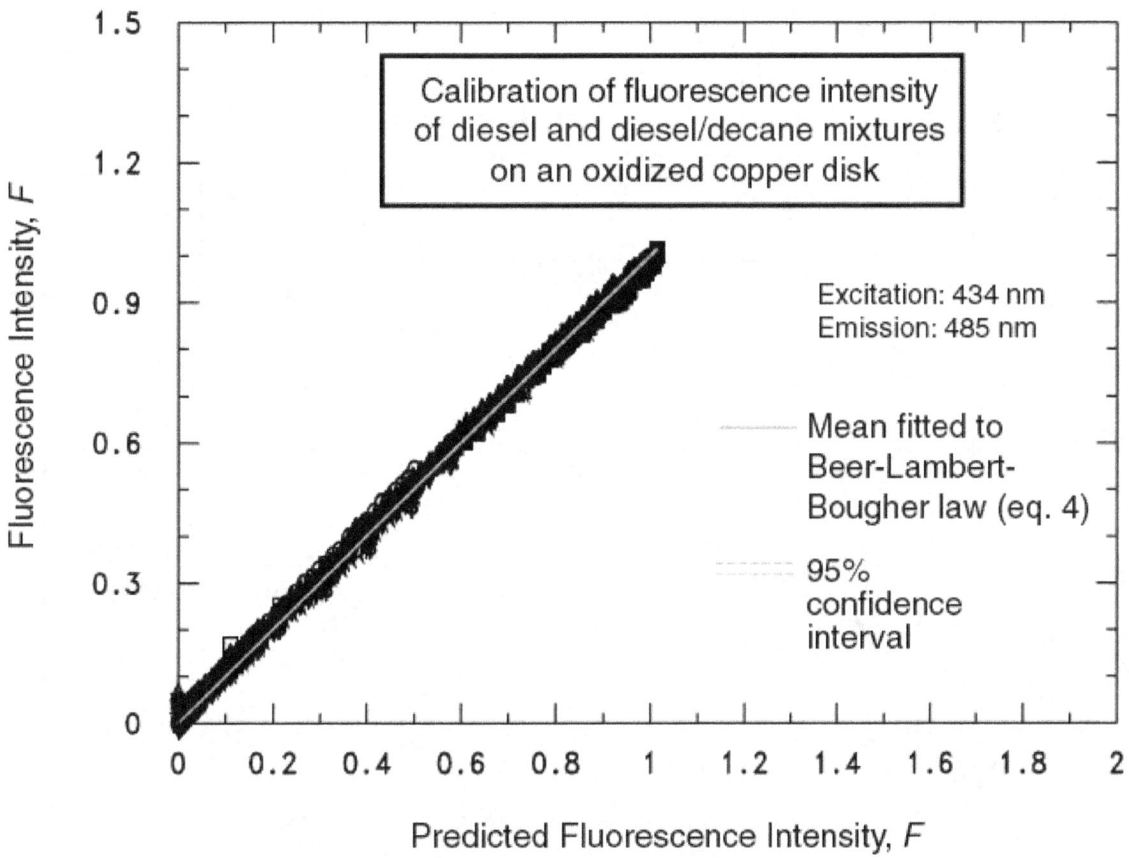

Fig. 6 Overall calibration of Beer-Lambert Bougher law for diesel on copper disk (Kedzierski, 2006)

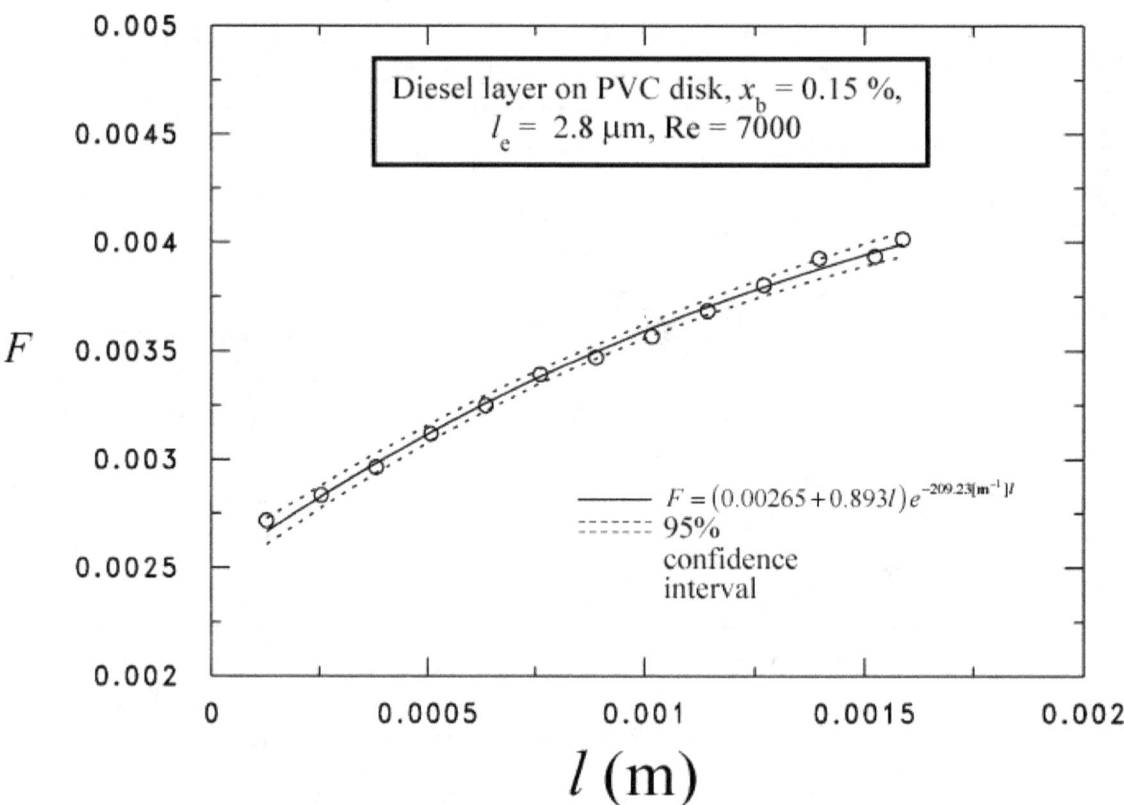

Fig. 7 Sample F vs. l fit for a PVC data set

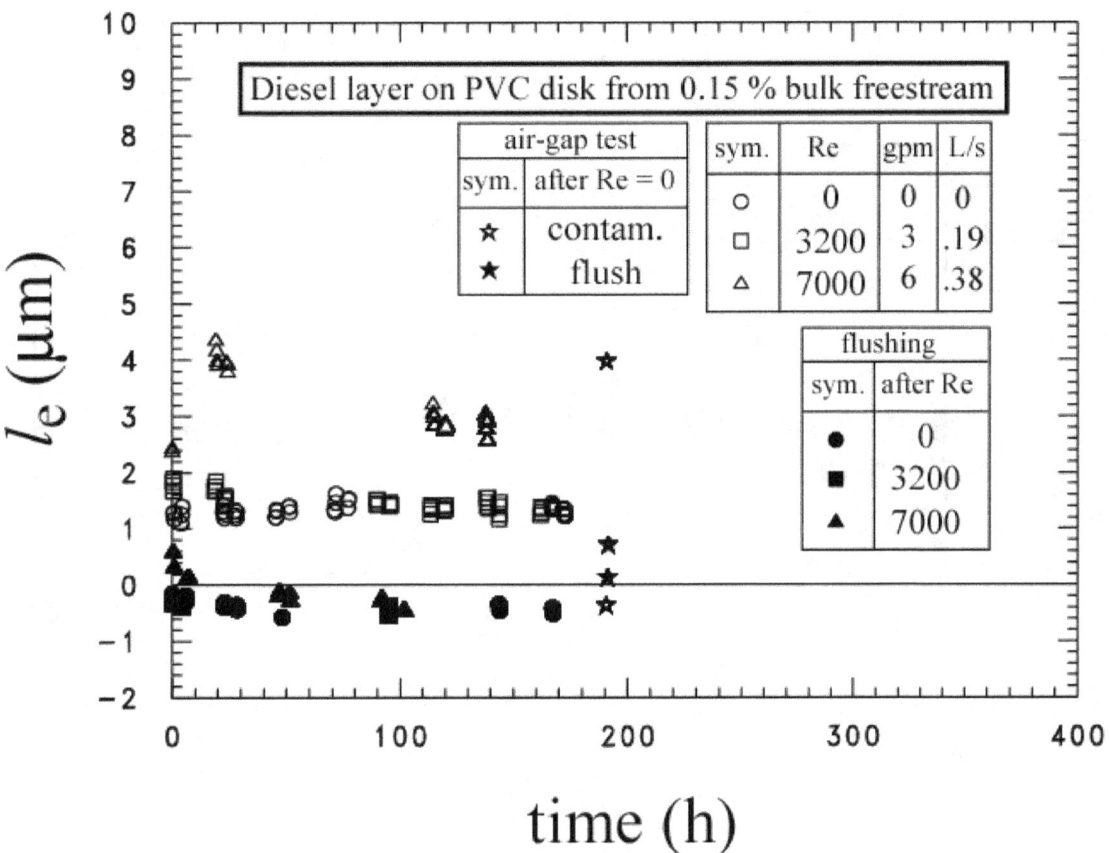

Fig. 8 Effect of exposure time and flow rate on thickness of the diesel excess layer for a 0.15 % bulk freestream mass fraction on a PVC disk

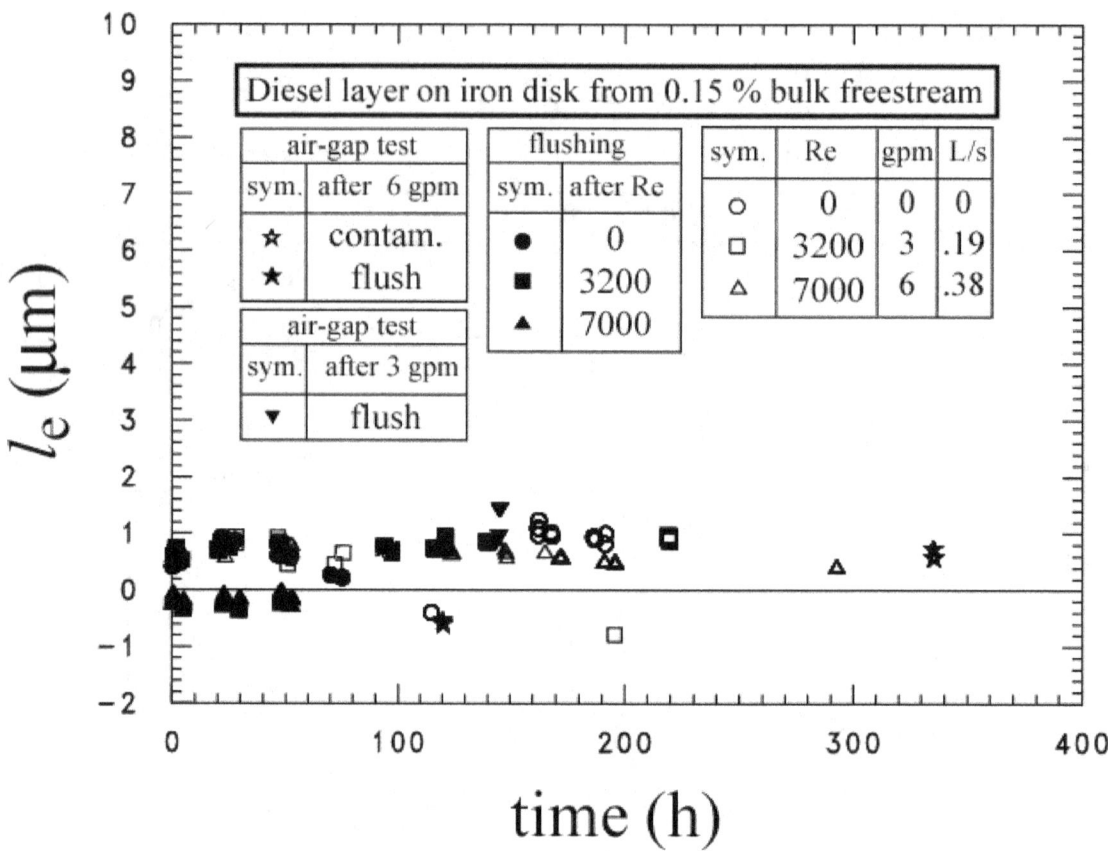

Fig. 9 Effect of exposure time and flow rate on thickness of the diesel excess layer for a 0.15 % bulk freestream mass fraction on an iron disk

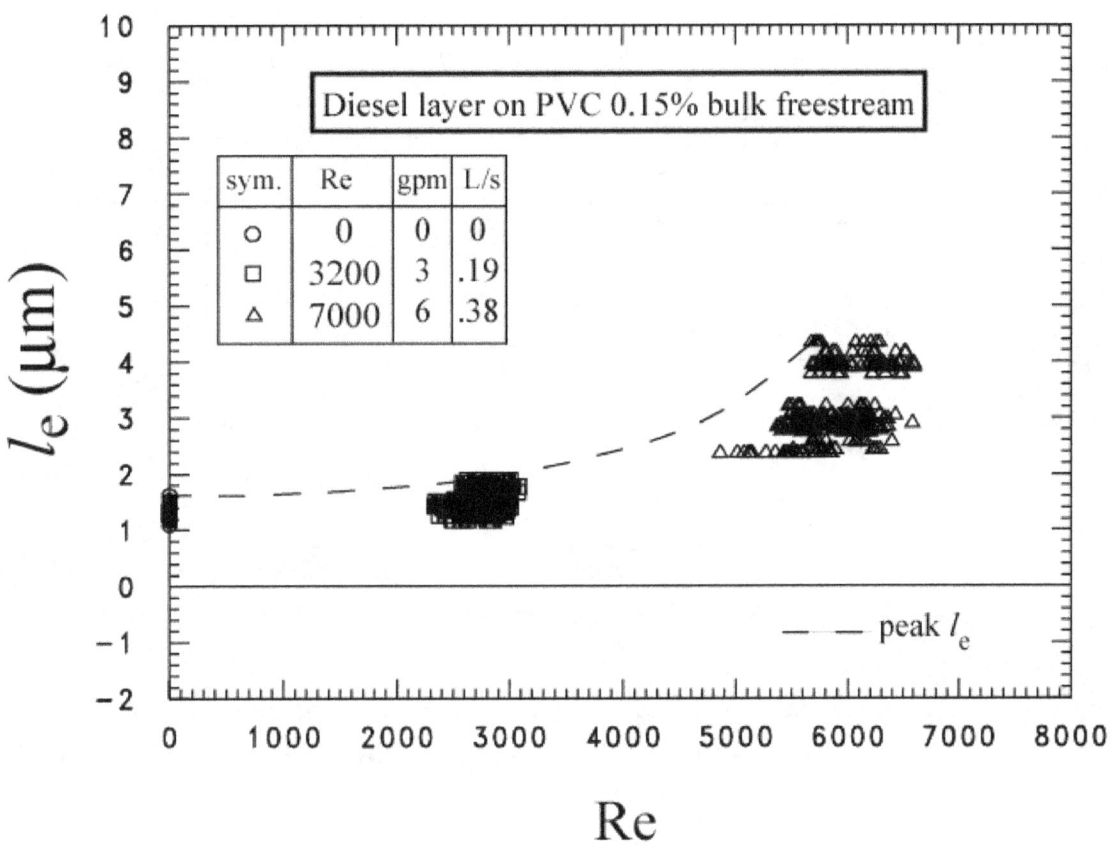

Fig. 10 Diesel excess layer thickness as a function of Re for PVC surface and water/diesel (99.85/0.15)

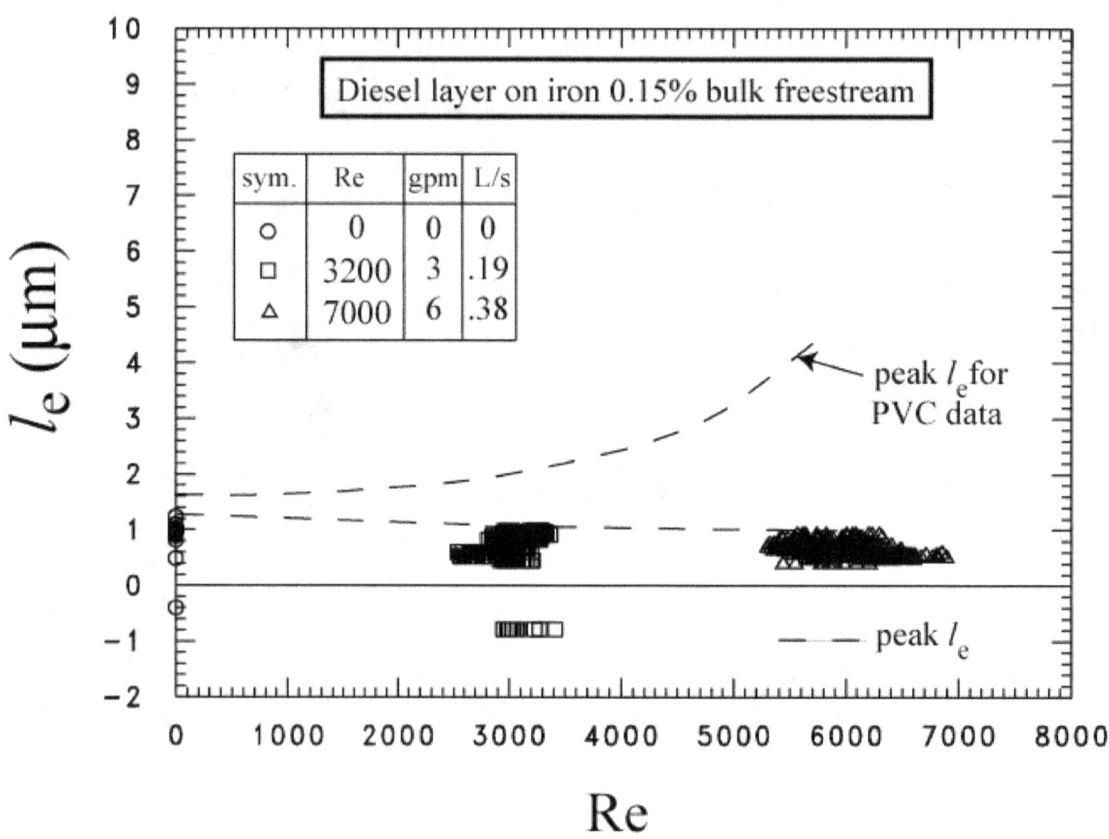

Fig. 11 Diesel excess layer thickness as a function of Re for iron surface and water/diesel (99.85/0.15)

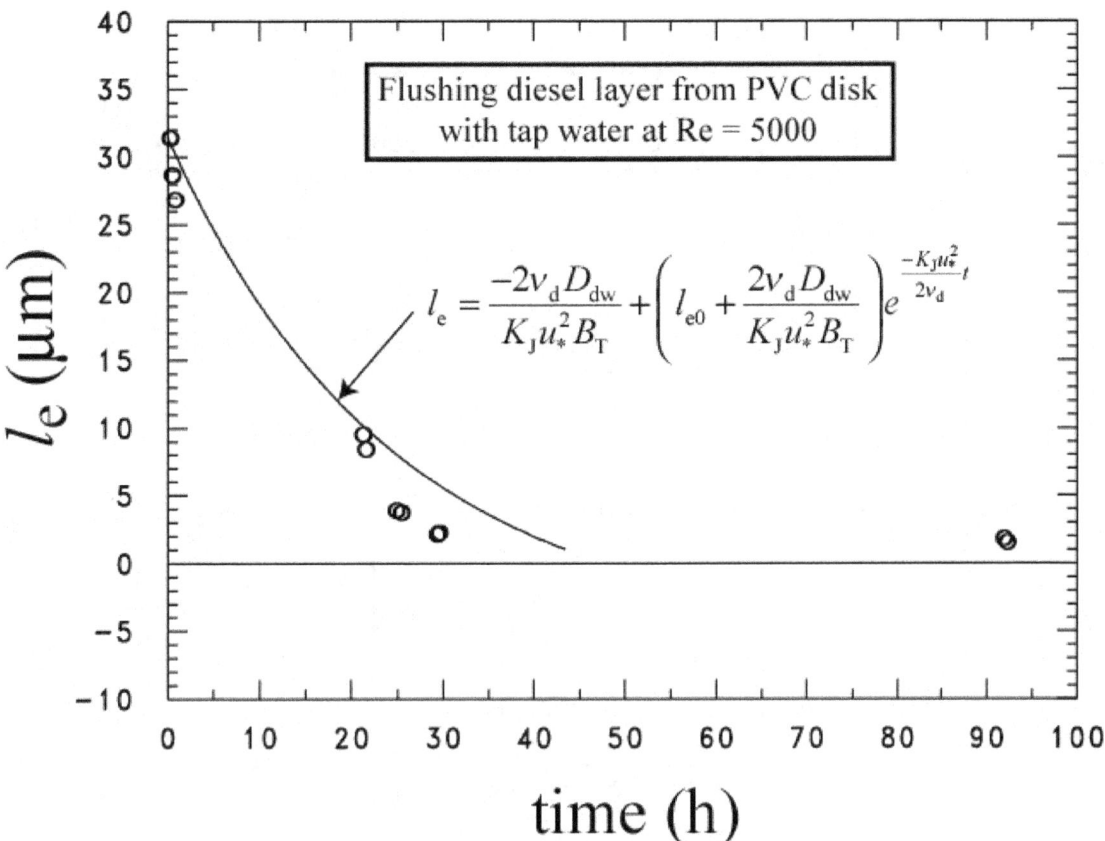

Fig. 12 Flushing measurements used to fit coefficients of model

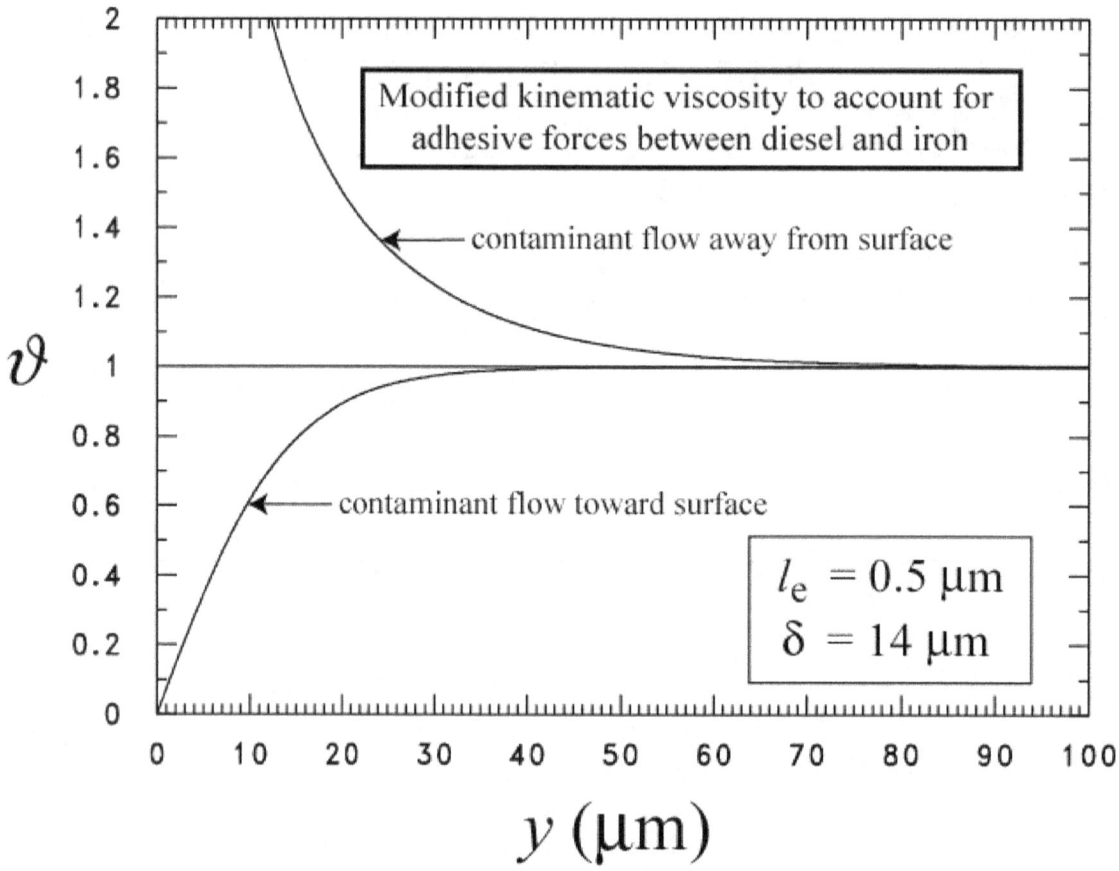

Fig. 13 Modified kinematic viscosity to account for adhesive forces between iron and diesel derived for two flow conditions (Re = 3200 and Re = 7000).

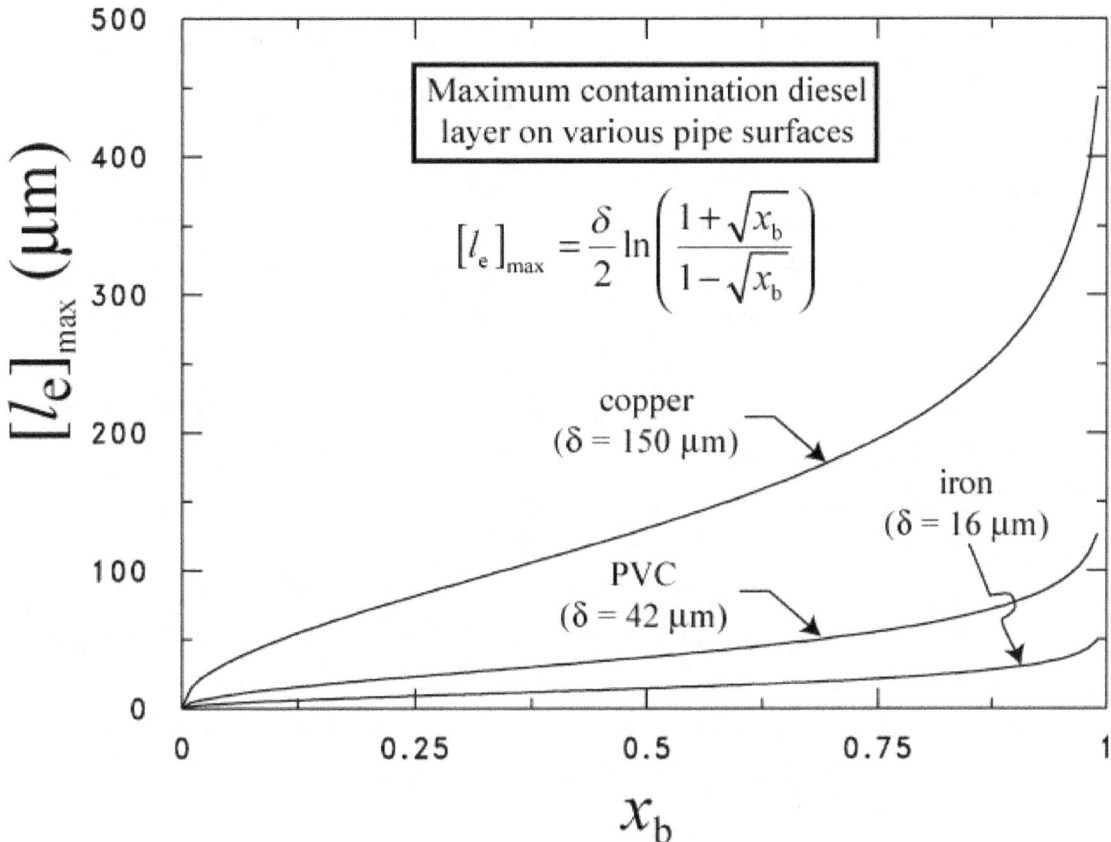

Fig. 14 Maximum contamination diesel layer on various pipe surfaces as predicted by eq. (22)

APPENDIX A: UNCERTAINTIES

Figure A.1 shows the relative (percent) uncertainty of the diesel excess layer thickness (U_{le}) as a function of l_e for the iron test surface. Measurements of l_e on the iron surface with uncertainties larger than 10 % were discarded. Of the retained measurement sets, the average uncertainty in l_e for the contamination and the flushing tests was approximately ± 8 %, and ± 4 % of l_e, respectively. Overall, the average uncertainty of l_e on an absolute basis for the iron surface contamination and flushing tests was approximately ± 0.06 µm and ± 0.02 µm, respectively.

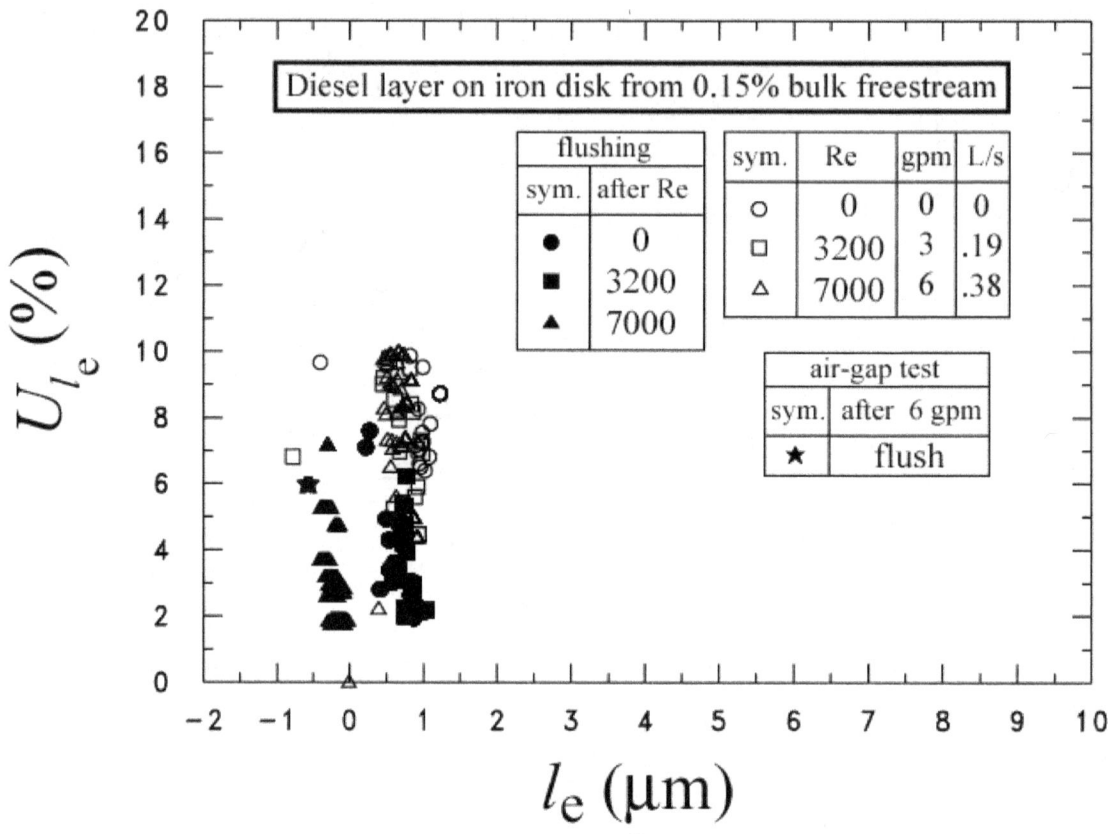

Fig. A.1 Relative uncertainty of l_e for 95 % confidence level for iron surface x_b = 0.15 %

Similarly, Fig. A.2 shows the relative (percent) uncertainty of the diesel excess layer thickness (U_{le}) as a function of l_e for the PVC test surface. For the PVC surface, contamination and flushing measurements l_e with uncertainties larger than 10 % and 20 %, respectively, were discarded. Of the retained measurement sets, the average uncertainty in l_e for the two lowest Re contamination tests and the Re = 7000 contamination tests was approximately ± 6 %, and ± 3 % of l_e, respectively. Overall, the average uncertainty of l_e for the measurements with the PVC surface for the contamination measurements was approximately ± 0.1 µm. The average uncertainty in l_e for the PVC flushing tests was approximately ± 17 % of l_e, which was approximately ± 0.05 µm.

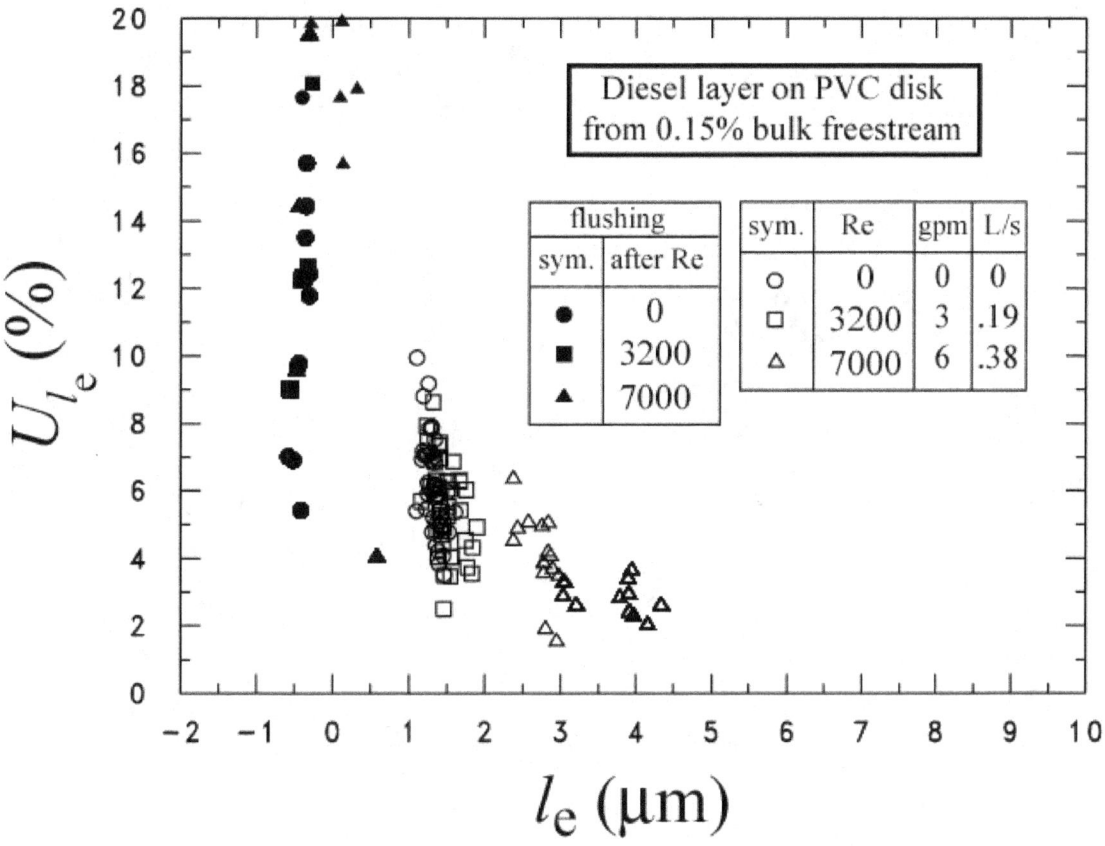

Fig. A.2 Relative uncertainty of l_e for 95 % confidence level for PVC surface x_b = 0.15 %

Figure A.3 shows the relative (percent) uncertainty of the bulk freestream diesel mass fraction (U_{xb}) as a function of x_b for the iron test surface. Of the retained measurement sets, the average uncertainty in x_b for the contamination tests was approximately ± 4 % of x_b. Overall, the average uncertainty of x_b on an absolute basis for the iron surface was approximately ± 0.006 % (± 60 ppm).

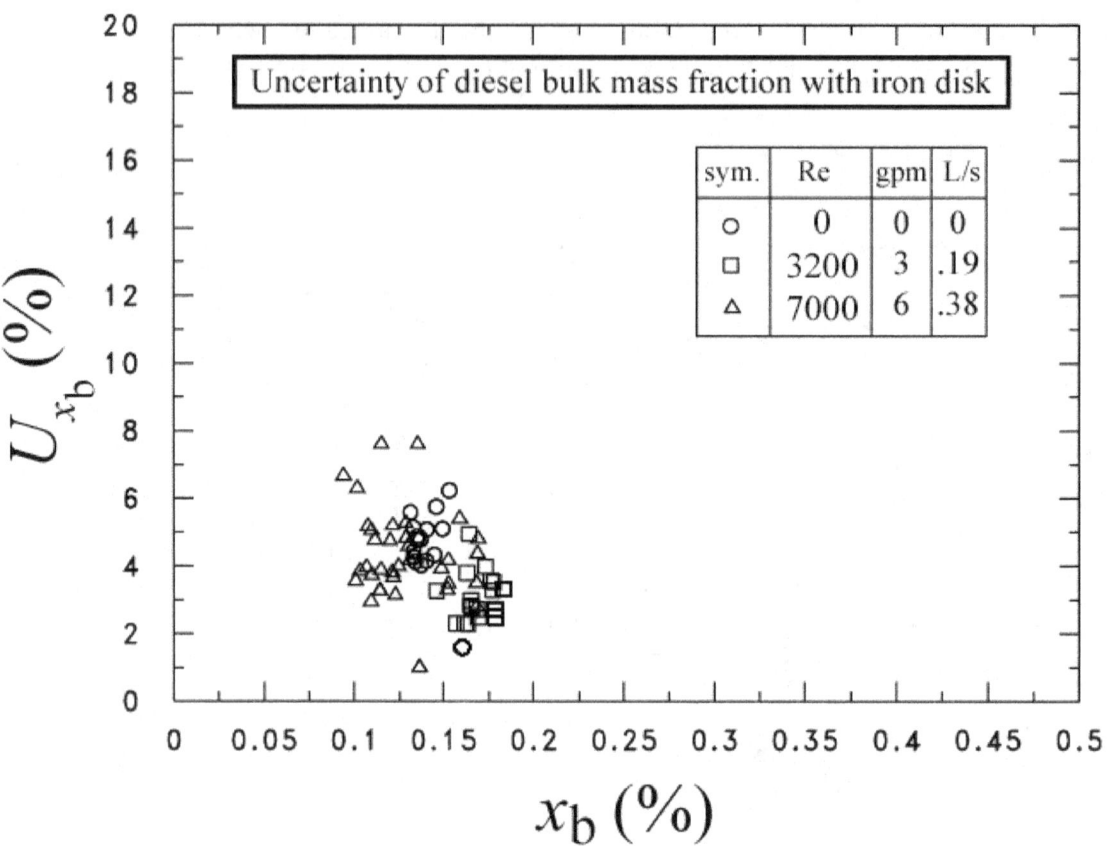

Fig. A.3 Relative uncertainty of x_b for 95 % confidence level for iron surface

Figure A.4 shows the relative (percent) uncertainty of the bulk freestream diesel mass fraction (U_{xb}) as a function of x_b for the PVC test surface. Of the retained measurement sets, the average uncertainty in x_b for the contamination tests was approximately ± 5 % of x_b. Overall, the average uncertainty of x_b on an absolute basis for the PVC surface was approximately ± 0.008 % (± 80 ppm).

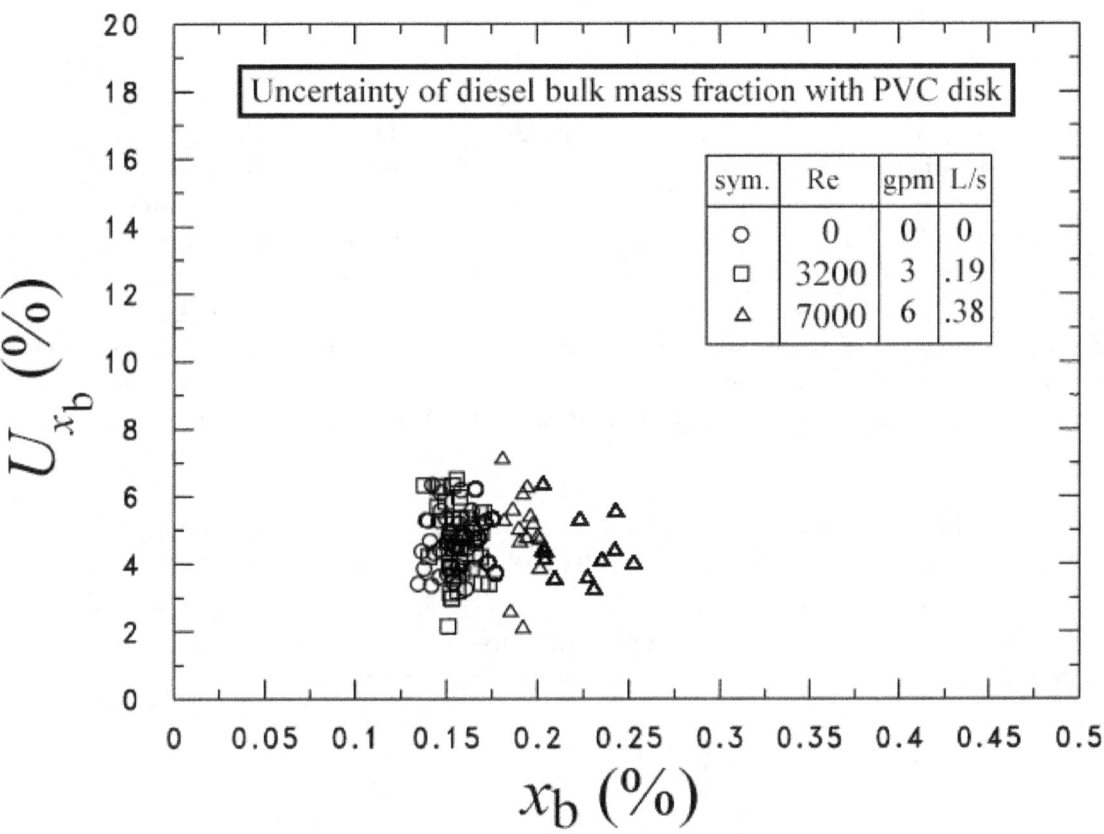

Fig. A.4 Relative uncertainty of x_b for 95 % confidence level for PVC surface

APPENDIX B: MEASUREMENT CORRECTION

This appendix outlines how two flushing tests and one contamination test were corrected to account for additional rust resulting when the iron surface was exposed to air during a one-time repair of the apparatus. The bottom of the quartz probe housing became ajar just prior to the flushing tests that were done after the Re = 3200 contamination tests. Due to exposure of air during the repair, rust developed on the iron test surface causing a reduction in the measured fluorescence intensity. Consequently, flushing tests after Re = 0 and the Re = 3200 contamination tests were corrected as were the soak contamination measurements.

The correction was accomplished by comparing repeated contamination measurements before repair and after repair for the same bulk mass fraction of diesel and the same Reynolds number. Figure B.1 shows the excess layer as a function of exposure time for Re = 7000 and $x_b = 0.0015$ for contamination tests before and after the test apparatus repair. The correction was devised to adjust the fluorescence intensity by a constant such that the average l_e of the before and after data sets were equal. Given that both data sets appeared to exhibit random trends with respect to time, it was assumed that a steady state between contaminant deposition and removal had been achieved for the entire data set. For this reason the entire data range for both data sets were used to obtain an average for each.

Because the oxidation of the iron had relatively ceased once it was re-immersed in the test water, the effect of the rust on the signal intensity was expected to be constant. In order to ensure that the adjustment to the intensity (ΔF_c) was a constant, the effect of the increased incident intensity with increased proximity to the surface was accounted for by the exponential term in the expression for the corrected fluorescence intensity (F_c):

$$F_c = F e^{209.23 [m^{-1}] l} + \Delta F_c \qquad (B.1)$$

The correction was found to be: $\Delta F_c = 0.001$ and the exponential term came from the calibration given in eq. (4). Once the intensity was corrected, F_c was used in place of F in the determination of l_e. The estimated uncertainty of l_e on an absolute basis for the corrected measurements was approximately ± 0.8 μm.

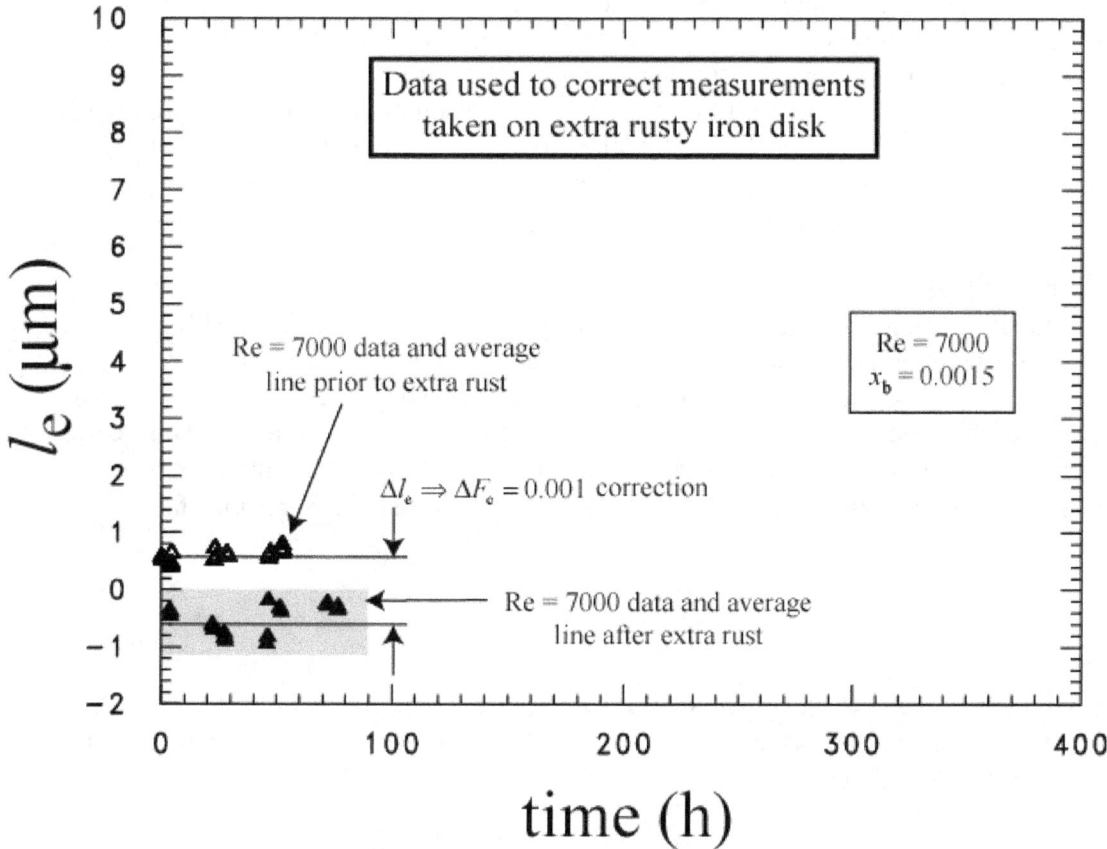

Fig. B.1 Basis for correction of iron surface measurements made after test apparatus repair

APPENDIX C: MODEL DEVELOPMENT

The derivations of the models for the prediction of the thickness of the contaminant excess layer on plumbing surfaces for the contamination and the flushing conditions are presented in this appendix. Each model assumes that transfer of mass to and from the surface occurs solely in a direction that is perpendicular to the surface. The flushing model predicts the thickness of the excess layer as a function of time, contaminant transport properties, and flushing Reynolds number. The contaminant model gives the maximum contaminant excess layer thickness that can occur for a given contaminant bulk mass fraction and surface affinity.

Flushing Model

The flushing model is based on the conservation of contaminant mass within the excess surface density layer. Because the excess layer is thin, it is approximated as stagnate in the axial direction such that the net motion of diesel is solely perpendicular from the surface in the, y-direction. From the corresponding continuity equation, the rate of change of change of mass (M) in the excess layer with respect to time (t) is equal to the mass flow of contaminant ($\dot{m}_d$) from the excess layer-flushing water interface:

$$\frac{\partial M}{\partial t} = \dot{m}_i - \dot{m}_o = -\dot{m}_d \qquad (C1)[5]$$

The mass of the excess layer can be represented by the product of the contaminant density (ρ_d), the surface area of the excess layer-water interface (A), and the thickness of the excess layer (l_e), i.e.:

$$M = \Gamma A = \rho_d A l_e \qquad (C2)$$

The diesel is transported from the surface via convection and diffusion. For turbulent flushing, convection of the diesel from the interface between the excess layer and the flushing water is caused by the pumping action of the fluctuating y-velocity component, $[v]_{y=l_e}$, that entrains diesel from the excess layer into the bulk water flow. Diffusion also occurs perpendicular to the surface because of the difference in diesel mass concentration between the excess layer ($\bar{c}_d$) and the water ($\bar{c}_w$). The following equation approximates the diffusion with an expression for equimolal diffusion (McCabe and Smith, 1976) that contains a diffusion coefficient (D_{wd}) and a transition depth (B_T) over which the mass concentration difference occurs. The total mass flux of diesel from the excess layer is the sum of the convective and the diffusive components:

$$\frac{\dot{m}_d}{A} = [v]_{y=l_e} \bar{c}_d + \frac{D_{wd}}{B_T}(\bar{c}_d - \bar{c}_w) \qquad (C3)$$

---

[5] Here $\dot{m}_i$ and $\dot{m}_o$ are the mass flow rates of contaminant entering and leaving the excess layer control volume, respectively. No flow enters the excess layer from the plumbing surface. Consequently, $\dot{m}_i$ is zero.

The effect of upstream diesel entrainment is assumed to not significantly increase the mass concentration of diesel in the flushing water. As a result, the c_w is approximated as zero. In addition, the diesel excess layer is neat diesel, i.e.:

$$\overline{c}_d = x_d \rho_d = \rho_d \tag{C4}$$

The average entrainment velocity (v) is approximated by half the magnitude of the maximum fluctuating component of the turbulent y-velocity (v'_{max}):

$$v = \frac{v'_{max}}{2} \tag{C5}$$

From the law of the wall, the axial velocity component in the viscous sublayer is (Tennekes and Lumley, 1982):

$$u = \frac{u_*^2}{\nu_d} y \propto u'_{max} \tag{C6}$$

where ν_d is the kinematic viscosity of the diesel, u'_{max} is the maximum fluctuating component of the turbulent axial velocity component, and u_* is the friction velocity given by Kays and Crawford (1980):

$$u_* = V\sqrt{0.039 \operatorname{Re}^{-0.25}} \tag{C7}$$

Where V is the bulk average velocity of the flushing water in the axial direction and Re is the Reynolds number based on the hydraulic diameter evaluated for the properties of the flushing water.

Kays and Crawford (1980) also show that the fluctuation turbulent velocity components may be related by a constant $K \equiv v'_{max} / u'_{max}$. If it is assumed that u'_{max} is proportional to the axial velocity, i.e., $u'_{max} \equiv Ju$, and the constant J is lumped into the constant K_J[6], v'_{max} can be written as:

$$[v]_{y=l_e} = \frac{[v'_{max}]_{y=l_e}}{2} = \frac{[K_J u_*^2 y / \nu_d]_{y=l_e}}{2} = \frac{K_J u_*^2 l_e}{2\nu} \tag{C8}$$

---

[6] $K_J \equiv KJ \equiv v'_{max} / u$

Substitution of eqs. (C2), (C3), (C4) and (C8) yields the following first order partial differential equation that governs the removal of the excess layer (l_e) during flushing:

$$\frac{\partial l_e}{\partial t} + \frac{K_j u_*^2 l_e}{2v_d} = -\frac{D_{dw}}{B_T} \tag{C9}$$

multiplying both sides of the differential equation by the integrating factor $e^{\frac{K_j u_*^2}{2v_d}t}$ gives:

$$e^{\frac{K_j u_*^2}{2v_d}t}\left(\frac{\partial l_e}{\partial t} + \frac{K_j u_*^2 l_e}{2v_d}\right) = -\frac{D_{dw}}{B_T} e^{\frac{K_j u_*^2}{2v_d}t} = \frac{\partial\left(l_e e^{\frac{K_j u_*^2}{2v_d}t}\right)}{\partial t} \tag{C10}$$

Separating and integrating the last two terms of the above equation and ensuring that $l_e = l_{e0}$ at $t = 0$ yields:

$$l_e = \frac{-2v_d D_{dw}}{K_j u_*^2 B_T} + \left(l_{e0} + \frac{2v_d D_{dw}}{K_j u_*^2 B_T}\right) e^{\frac{-K_j u_*^2}{2v_d}t} \tag{C11}$$

The required time to flush l_{e0} to $l_e = 0$ can be determined by setting l_e to zero in the above equation and solving for the elapsed time t:

$$t = \frac{-2v_d}{K_j u_*^2} \ln\left[\frac{1}{\frac{K_j u_*^2 l_{e0} B_T}{2v_d D_{dw}} + 1}\right] \tag{C12}$$

Contamination Model
For the contamination case where a balance between deposition of contaminant on the surface and removal of the contaminant from the surface must be achieved, surface adhesion requires that a distinction be made between the velocity of the contaminant toward the surface (v_i) and that away from it (v_o). Very near the wall there is an additional resistance to flow away from the surface as caused by the attraction of diesel molecules to the molecules of the pipe surface. Likewise, for the same region near the pipe, the attractive forces induce a reduction in the resistance of flow toward the surface. Considering that it is flow that is being modeled, a simply way to approximate this behavior is via a modified viscosity. For example, the entrainment velocity (evaluated using the properties of the contaminated flow) given in eq. (C8) for flow approaching the pipe surface becomes:

$$[v_i]_{y=l_e} = \frac{K_j u_*^2 l_e}{2v_d \tanh\left(\dfrac{l_e}{\delta}\right)} \tag{C13}$$

Here, the modified kinematic viscosity is less than the constant property viscosity (v_d) for distances from the surface less than the penetration depth, δ. The magnitude of the penetration depth is determined by the affinity of the contaminant for the pipe surface. Consequently, each contaminant\pipe combination will have its own value of δ. The hyperbolic tangent was chosen for its simplicity and because it closely matched what was believed to be the required relationship with respect to l_e.

Likewise, surface adhesion acts to deter the flow of contaminant from the wall according to an assumed cotangent relationship with respect to l_e. Here the leaving velocity is approximated as a local increase in viscosity above the constant property value as the pipe surface is approached for distances less than δ:

$$[v_o]_{y=l_e} = \frac{K_j u_*^2 l_e}{2v_d \coth\left(\dfrac{l_e}{\delta}\right)} \tag{C14}$$

For the case where diffusion is negligible, the balance between the deposition and removal of contaminant at the wall becomes:

$$\frac{\partial M}{\partial t} = \dot{m}_i - \dot{m}_o = \left([v_i]_{y=l_e} x_b \rho_d - [v_o]_{y=l_e} \rho_d\right) A \tag{C15}$$

Substitution of M from eq. (C2) reduces the above mass balance to:

$$\frac{\partial l_e}{\partial t} = [v_i]_{y=l_e} x_b - [v_o]_{y=l_e} \tag{C16}$$

Substitution of v_i and v_o into eq. (C16) gives the following first order differential equation for surface contamination from a bulk flow with a contaminant mass fraction of x_b:

$$\frac{\partial l_e}{\partial t} = \left(\frac{K_j u_*^2 l_e}{2v_d}\right)\left(\frac{x_b}{\tanh\left(\dfrac{l_e}{\delta}\right)} - \frac{1}{\coth\left(\dfrac{l_e}{\delta}\right)}\right) \tag{C17}$$

Considering that the exposure time a pipe surface to a contaminant may not always be known. A conservative decontamination response would be to flush for the maximum contaminant film thickness (excess layer) for the given concentration of contaminant in the flow. The maximum contaminant excess layer, which occurs at steady state where the deposition balances the removal at the wall, can be determined by setting the partial derivative of the excess layer with respect to time to zero in eq. (C17) and solving the resulting equation for l_e, yielding:

$$[l_e]_{max} = \frac{\delta}{2} \ln\left(\frac{1+\sqrt{x_b}}{1-\sqrt{x_b}}\right) \rightarrow for \rightarrow 0 < x_b < 1 \qquad (C18)$$

APPENDIX D: FLUORESCENCE OF TOLUENE

This appendix presents the fluorescence of a potential next contaminant to study using the fluorescence measurement technique presented in this report. Figure D.1 schematically shows the test setup to measure the fluorescence emission of toluene on PVC, copper, and iron surfaces. Each emission was measured with the spectrofluorometer instrumented with two fiber optical bundles orientated nearly perpendicular to each other at the test surface. The excitation bundle was orientated at a slight angle from the horizontal with respect to the toluene film that was on the test surface. The emission bundle was perpendicular to the test film and surface.

The toluene film was confined to each test surface by an open-ended quartz box that was sealed around the 10 mm x 30 mm outside perimeter of the base with modeling clay. In order to achieve a repeatable film thickness of approximately 1 mm, single drops of toluene were dropped onto the surface until it was fully wetted.

Each emission spectrum was produced with a constant excitation wavelength of 355 nm. Figure D.2 shows the fluorescence emission of toluene on three pipe surfaces: PVC, copper, and iron. Figure D.2 shows that the surface materials do not significantly affect the emission spectrum. Each emission spectrum has approximately the same shape and the same peak wavelength. The tests show that maximum signal intensity for toluene is found between 375 nm and 425 nm and that toluene is a potential contaminant candidate that can be detected with the measurement technique that is presented in this report. The uncertainty of the mean intensity and wavelength measurement for this particular experimental setup was approximately ± 20 %, and ± 2 nm, respectively.

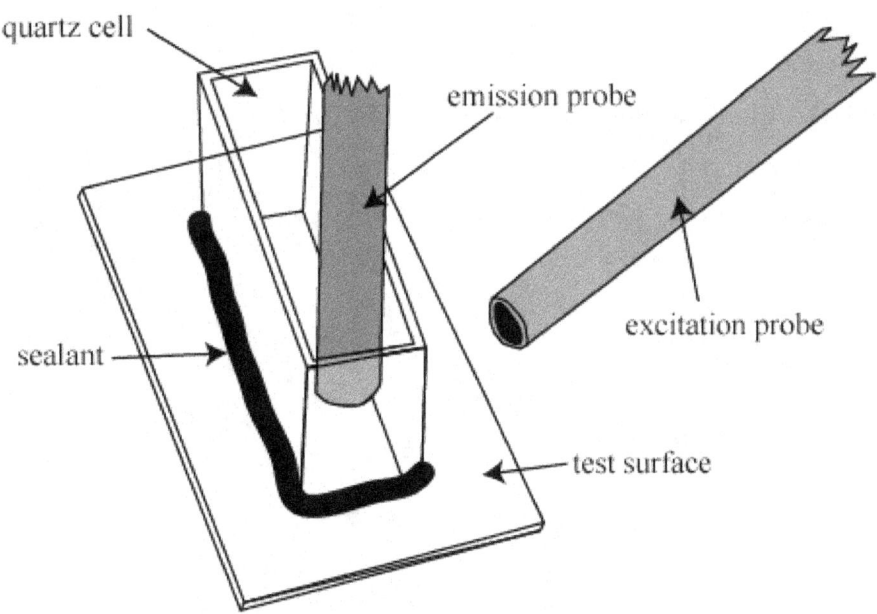

Fig. D.1 Schematic of experimental setup to measure fluorescence emission of toluene on three pipe surfaces: PVC, copper, and iron

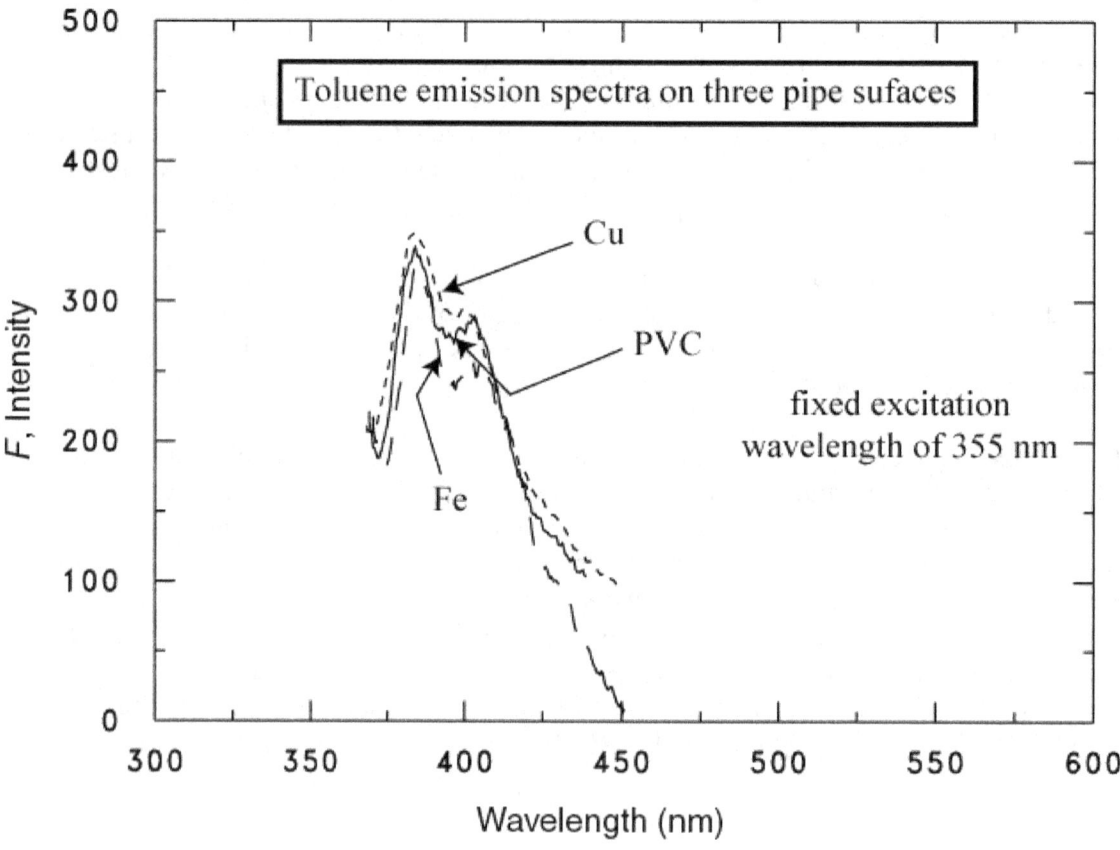

Fig. D.2 Fluorescence emission of toluene on three pipe surfaces: PVC, copper, and iron

www.ingramcontent.com/pod-product-compliance
Lightning Source LLC
Chambersburg PA
CBHW081905170526
45167CB00007B/3155